Communications in Computer and Information Science 1823

Rationale

The CCIS series is devoted to the publication of proceedings of computer science conferences. Its aim is to efficiently disseminate original research results in informatics in printed and electronic form. While the focus is on publication of peer-reviewed full papers presenting mature work, inclusion of reviewed short papers reporting on work in progress is welcome, too. Besides globally relevant meetings with internationally representative program committees guaranteeing a strict peer-reviewing and paper selection process, conferences run by societies or of high regional or national relevance are also considered for publication.

Topics

The topical scope of CCIS spans the entire spectrum of informatics ranging from foundational topics in the theory of computing to information and communications science and technology and a broad variety of interdisciplinary application fields.

Information for Volume Editors and Authors

Publication in CCIS is free of charge. No royalties are paid, however, we offer registered conference participants temporary free access to the online version of the conference proceedings on SpringerLink (http://link.springer.com) by means of an http referrer from the conference website and/or a number of complimentary printed copies, as specified in the official acceptance email of the event.

CCIS proceedings can be published in time for distribution at conferences or as post-proceedings, and delivered in the form of printed books and/or electronically as USBs and/or e-content licenses for accessing proceedings at SpringerLink. Furthermore, CCIS proceedings are included in the CCIS electronic book series hosted in the SpringerLink digital library at http://link.springer.com/bookseries/7899. Conferences publishing in CCIS are allowed to use Online Conference Service (OCS) for managing the whole proceedings lifecycle (from submission and reviewing to preparing for publication) free of charge.

Publication process

The language of publication is exclusively English. Authors publishing in CCIS have to sign the Springer CCIS copyright transfer form, however, they are free to use their material published in CCIS for substantially changed, more elaborate subsequent publications elsewhere. For the preparation of the camera-ready papers/files, authors have to strictly adhere to the Springer CCIS Authors' Instructions and are strongly encouraged to use the CCIS LaTeX style files or templates.

Abstracting/Indexing

CCIS is abstracted/indexed in DBLP, Google Scholar, EI-Compendex, Mathematical Reviews, SCImago, Scopus. CCIS volumes are also submitted for the inclusion in ISI Proceedings.

How to start

To start the evaluation of your proposal for inclusion in the CCIS series, please send an e-mail to ccis@springer.com.

Filippo Neri · Ke-Lin Du ·
Vijayakumar Varadarajan ·
Angel-Antonio San-Blas · Zhiyu Jiang
Editors

Computer and Communication Engineering

Third International Conference, CCCE 2023
Stockholm, Sweden, March 10–12, 2023
Revised Selected Papers

 Springer

Editors
Filippo Neri 🄳
University of Naples Federico II
Naples, Italy

Ke-Lin Du 🄳
Concordia University
Montreal, QC, Canada

Vijayakumar Varadarajan 🄳
The University of New South Wales
Sydney, NSW, Australia

Angel-Antonio San-Blas 🄳
Miguel Hernández University of Elche
Elche, Spain

Zhiyu Jiang 🄳
Northwestern Polytechnical University
Xi'an, China

ISSN 1865-0929   ISSN 1865-0937 (electronic)
Communications in Computer and Information Science
ISBN 978-3-031-35298-0   ISBN 978-3-031-35299-7 (eBook)
https://doi.org/10.1007/978-3-031-35299-7

This Springer imprint is published by the registered company Springer Nature Switzerland AG
The registered company address is: Gewerbestrasse 11, 6330 Cham, Switzerland

Preface

The 3rd International Conference on Computer and Communication Engineering (CCCE 2023) was successfully held in Stockholm, Sweden during March 10–12, 2023. The objective of CCCE 2023 was to provide an interactive forum for presentations and discussions in computer and communication engineering and related fields.

CCCE 2023 provided a scientific platform for both local and international scientists, engineers and technologists who work in all areas of computer and communication engineering to share much more than technical interests – among other things, culture and history. In addition to the contributed papers, internationally known experts from several countries were also invited to deliver keynote and invited speeches at CCCE 2023. They are Nikola Kasabov (Auckland University of Technology, New Zealand), Angrisani Leopoldo (Università degli Studi di Napoli Federico II, Italy), Arumugam Nallanathan (Queen Mary University of London, UK). The whole conference was held online in Zoom; we also had tests for each presenter in advance to ensure the successful delivery of the conference. For those who had internet problems, a pre-recorded video presentation was accepted as an alternative. Meanwhile, the whole conference was recorded for conference backup only.

The final proceedings of CCCE 2023 include 18 papers, which were selected after a thorough reviewing process. We are grateful to the authors and the participants, as they are the key components for providing an ideal forum to exchange results and ideas. We also would like to express our sincere gratitude to our colleagues in the disciplines who have kindly volunteered in the review process. Special thanks go to the conference organizers and session chairs for their devotion in developing a strong technical program and enjoyable social events. It is our hope that the proceedings of CCCE 2023 will result in increased collaboration among researchers in the fields and will help to facilitate exploration of novel cross-disciplinary ideas.

The international conference on computer and communication engineering creates a good opportunity for researchers, scientists, and engineers from different parts of the world to share their latest knowledge and findings on a variety of topics, including Image Analysis and Methods, Network Models and Function Analysis of Mobile Networks, System Security Estimation and Analysis of Data Networks, and AI -based System Models and Algorithms.

On behalf of the organizing committee, we hope that you had a fruitful meeting at this conference. We will continue to organize this conference in the future to provide a highly effective platform for further exchange of new knowledge and perhaps potential collaboration in the research areas of computer and communication engineering.

Warmest regards,

March 2023   Filippo Neri

Conference Committees

Advisory Committees

Arumugam Nallanathan (IEEE Fellow)	Queen Mary University of London, UK
Angrisani Leopoldo (IEEE Fellow)	Università degli Studi di Napoli Federico II, Italy
Nikola Kasabov (IEEE Fellow)	Auckland University of Technology, New Zealand
Patrick Siarry	Université Paris-Est Créteil, France

Conference General Chair

Filippo Neri	University of Naples "Federico II", Italy

Conference General Co-chair

Ning Xiong	Mälardalen University, Sweden

Program Chairs

Ke-Lin Du	Concordia University, Canada
Vijayakumar Varadarajan	University of New South Wales, Australia

Program Co-chairs

Angel-Antonio San-Blas	Miguel Hernández University of Elche, Spain
Zhiyu Jiang	Northwestern Polytechnical University, China

Technical Committees

Yew Kee Wong	BASIS International School Guangzhou, China
Christoph Lange	HTW Berlin University of Applied Sciences, Germany
Toufik Bakir	University of Burgundy, France
Zhiliang Qin	Shandong University, China
Gábor Földes	Budapest University of Technology and Economics, Hungary
Smain Femmam	Haute-Alsace University, France
Paolo Garza	Polytechnic University of Turin, Italy
Shihab A. Hameed	International Islamic University Malaysia, Malaysia
Călin Ciufudean	Stefan cel Mare University of Suceava, Romania
Petr Hajek	Academy of Sciences of the Czech Republic, Czech Republic
Libor Pekař	Tomas Bata University in Zlín, Czech Republic
Pius Adewale Owolawi	Tshwane University of Technology, South Africa
Nitikarn Nimsuk	Thammasat University, Thailand
Mandeep Singh Jit Singh	Universiti Kebangsaan Malaysia, Malaysia
Cuong Pham-Quoc	Ho Chi Minh City University of Technology, Vietnam
George Petrea	"Dunarea de Jos" University of Galati, Romania
Tony Spiteri Staines	University of Malta, Malta
Noor Zaman Jhanjhi	Taylor's University, Malaysia
Toe Wai Tun	Pyinoolwin Science and Technology Research Center, Myanmar
Manoj Gupta	JECRC University, India
K. Vinoth Kumar	SSM Institute of Engineering and Technology, India
Wahyu Pamungkas	Institut Teknologi Telkom Purwokerto, Indonesia
Mayukha Pal	ABB Ltd., India
Parameshachari B.D.	GSSS Institute of Engineering and Technology for Women, India

Supporters

Contents

AI-Based System Model and Algorithm

Image Analysis and Method

Enhanced Acoustic Noise Reduction Techniques for Magnetic Resonance Imaging System

I. Juvanna[1]([⊠]), Uppu Ramachandraiah[2], and G. Muthukumaran[1]

[1] Hindustan Institute of Technology and Science, Chennai, India
juvanna@gmail.com, gmkumaran@hindustanuniv.ac.in
[2] SRM Institute of Science and Technology, Chennai, India

Abstract. Magnetic Resonance Imaging is a diagnostic tool meant for scanning organs and structures inside the body. The undesirable effect of MRI machine is the significant high level of acoustic noise produced at the time of scanning, which creates more negative effects. Therefore there is a necessity to reduce this noise level. Various solutions, including software and hardware upgrades, can be utilized to address this issue. The suggested model includes an active noise control (ANC) system for pre-recorded sound from a MRI scanner. The modified ANC with Filtered Least Mean Square Algorithm (FxLMS) technique aids in reduction of noise level produced in the MRI scanner. Then the performance of the system can be analyzed by three different cases such as insulation of the chamber by glasswool, performance of noise level reduction in the presence of static and dynamic magnetic field, and analysis of noise level reduction by employing multiple microphones with this modified ANC system. The sound pressure level (SPL) is measured with and without ANC system. From the result analysis, it is found that the ANC system performance is enhanced to significant level of 24 dB noise reduction with glasswool insulation and 25 dB noise reduction with multiple microphones, while the static and dynamic magnetic field does not affect the system performance.

Keyword: Active Noise Control (ANC) · Magnetic Resonance Imaging (MRI) · Least Mean Square (LMS) · Filtered x Least Mean Square (fxLMS)

1 Introduction

In recent time medical field introduce numerous features and technologies for recognizing the health of human and diagnose several disease to reduce the death rate. In order to enhance the treatment process, medical imaging techniques are utilized (Hussain et al., 2017; Sartoretti et al., 2020). Some of the mostly used imaging approaches are Positron Emission Tomography (PET), Magnetic Resonance Imaging (MRI), Computed Tomography (CT), X-ray and Function Magnetic Resonance Imaging (FMRI), etc. (Li et al., 2006). In the field of modern medicine, imaging techniques are used to detect the appropriate stage of disease and to find the abnormalities in body, which provides a clear idea to doctor about the disease for early detection to reduce the death rate (Lee et al., 2017;

© The Author(s), under exclusive license to Springer Nature Switzerland AG 2023
F. Neri et al. (Eds.): CCCE 2023, CCIS 1823, pp. 3–15, 2023.
https://doi.org/10.1007/978-3-031-35299-7_1

Lu et al., 2021). Most widely used imaging technique is MRI, which generates a detailed three dimensional view of anatomical images by using powerful magnetic radio waves and magnetic fields.

MRI is a non-invasive imaging method, used to detect the exact condition of affected area inside the human body like abdomen, pelvis and chest, etc. Now a day's MRI imaging technique is used in both medical and research fields, which maps the structure of body and function (Meng and Chen 2020). However MRI imaging being used widely, it also produces a high level of unwanted noise called acoustic noise (Chang et al. 2013; Luo et al. 2017). In MRI, the acoustic noise arises during the continuous changes of current inside the gradient coil. This results in the Lorentz force interaction among conductors and magnetic field to produce vibration with high level of noise which harms the patient (Yamashiro et al. 2019; Furlong et al. 2021). Acoustic noises in MRI affects all age people, moreover it disconnects the interaction among the scanning technician and the patient.

Various techniques are introduced in reducing the unwanted noise of MRI, but it only minimize the coil assembly vibration, changing the design of gradient coil, generating a silent MRI pulse sequence, provide passive ear for noise prevention, etc. (Rondinoni et al. 2013; Kuo et al. 2018). Acoustic noise affects the patient by the rate of slew, strength of gradient, etc. To tackle all these problems raised in field of modern medicine, the Active Noise Control (ANC) is introduced (Roozen et al. 2008; Hennel 2001). ANC is also called as active noise reduction, noise cancellation, which is used to decrease the unnecessary noise. ANC utilize the microphone to extract the less amount of frequency noise for neutralizing the unwanted noise before it affects the patient (Rudd 2012; Kumamoto 2010). ANC method is widely used in many industrial fields and applications to remove the acoustic noise.

In MRI the acoustic noise is identified by ANC method, which requires two things for optimal noise reduction (McJury et al. 1997; Muto et al. 2014). The correct actuation method is detected first for unique magnetic environment. Then it requires an MRI compatible speaker (Pla 1995) for managing the wave of acoustic noise. Secondly an appropriate controller is developed for reducing the level of sound pressure efficiently (Ardekani and Abdulla 2014). In this work the acoustic noise produced by MRI machine is reduced efficiently by using the ANC method.

2   Related Works

Yamashiro et al., (2019) proposed an evaluation technique for MRI acoustic noise reduction technology by controlling the magnetic gradient waveform. In MRI systems, ComforTone is a noise - reducing technique, which reduces acoustic noise (AN) by altering pulse sequences. While ComforTone is used to tackle AN issues for those who are exposed to MRI noise, the associated technical details have yet to be disclosed, and its benefits on AN reduction are unknown.

Huang et al. (2020) proposed a noise control technique for headphones. In this work an active noise control technique is developed to reduce the level of acoustic noise for headphone. Active noise control is used in proposed approach to reduce the unwanted noises generated from the environment. In this work ANC headphones utilize

TMS320C7613to produce an anti-noise signal as the computing core. The audio signals are integrated by using anti-noise signal, which used to reduce the unwanted noise produced from environment. The experimental results of proposed method obtain high rate of bandwidth of noise reduction in headphones.

Lee et al. (2017) suggests an active noise control method for reducing acoustic noise in MRI. The unwanted noise produced in 3T MRI is measured and recognized by using time frequency domain, which includes higher harmonics of fundamental frequency and Fourier coefficients of all peaks. By assuming the noise is periodic for smaller duration the control signal is identified by sum of sinusoidal signal developed by Fourier coefficient. The experimental result of proposed approach obtained better performance by reducing the 35 dB in frequency range of interest.

Hasegawa et al. (2018) proposed a system based on psychoacoustic characteristics ANC system. This method incorporates noise weighting into the standard ANC framework. Both MRI noise and noise generated at random have been shown to be beneficial. The basic ANC architecture is combined with noise weighting in this ANC technique. Through numerous experimental findings and a subjective evaluation, the efficacy of the ANC system is examined.

Furlong et al. (2021) proposed a technique to test the possibility of active noise from MRI systems, researchers employed control generators outside the area, as MRI scanners are responsive to metallic components. With a 180° phase difference, the time reversal (TR) invert filtering approach is shown to be comparable to ANC. The transmission of a noise-cancelling message from a farther distance to a preferred place (such as a patient's ears) is the subject of this research. In order to establish the performance of a remote distribution system, a parameterization study is done to examine frequency dependency and signal length. The effectiveness of ANC in reducing MRI noise was demonstrated using MRI noise recordings.

Kang and Xia (2021) proposed an acoustic noise control method for MRI using gradients coil system process. A simple momentum equation is inserted into the classic gradients coil design concept utilising a finite-difference-based approach, which may integrate the stream aim and coil dislocation generated by fast gradient shifting. To demonstrate the suggested design approach, a three-dimensional slant gradients coils is provided as a design sample. Acoustic models are also built to approximate the sound pressure level (SPL).

3   Proposed Methodology

Active Noise Control: The frequencies as from sound transmitter and the secondary (loudspeaker) transmitter interact in a disrupting way at the error microphone. As illustrated in Fig. 1, ANC is the reduction of acoustic noise by generating a zone of destructive interference at a specific location using an anti-phase acoustic wave. Despite the fact that the concept of ANC was created in the 1930s, today's access to low-cost digital signal processing (DSP) equipment and the development of reliable DSP procedures has resulted in a wide range of applications.

The FxLMS algorithm can theoretically be used to tweak the ANC system to deliver the appropriate response. In the typical ANC system, the FxLMS algorithm is presented in Fig. 1. The main speaker and the acoustic plant's mathematical transfer function are represented by plant model $P(z)$, the undesirable noise sample is represented by input signal k^{th}, and residual noise is represented by $e(k)$. Both signals are measured using microphones. An anti-noise speaker is powered by the FxLMS N^{th} output signal, which cancels out unwanted noise. The error path, represented by $S(z)$, is the acoustic path from the anti-noise speaker to the error microphone. The FxLMS method uses weighting parameters to update the order ANC. As a result, the residual noise $e(k)$ can be written as

$$r(x) = A(x)L(y)M(y) + b(x)M(y) \tag{1}$$

where $B(x) = C(x)^T A(x)$,
the weighting parameter is $C(x) = [c_0(x), c_1(x), \ldots, c_{N-1}(x)]^T$.
and the input noise is $A(x) = [a(x), a(x-1), \ldots, a(x-N+1)]^T$.

The associated gradient estimation and weight modifications are written as,

$$\Delta_{1k} = \frac{\partial r^2(x)}{\partial C(x)} = 2r(x)[A(x)M(y)] \tag{2}$$

The adaptation process is expressed as shown

$$C(x+1) = C(x) - \mu\Delta_{1k} = C(x) - 2\mu_1 r(x)[A(x)M(y)] \tag{3}$$

where μ shows the learning rate. The FxLMS algorithm in Eq. (3) contains actual of $A(x)M(y)$, unlike that in the conventional least mean squares (LMS) algorithm in (4)

$$C(x+1) = c(x) - 2\mu_1 r(x)A(x) \tag{4}$$

To implement the ANC system, the reference signal $A(x)$ must be filtered by $\hat{M}(y)$, which is obtained by identifying $M(y)$.

The adjusting weights approach is also known as the steepest descent method. This method uses an iterative approach that ensures the quadratic performance index $r(x)$ is reduced in respect to the filter weight $C(x)$. The converge of adaptive filter will produce the minimal solution as a result of the gradient estimation theorem, resulting in residual noise $r(x) \to 0$ as $x \to \infty$.

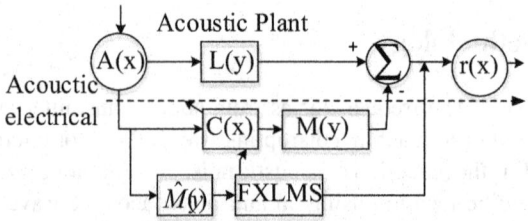

Fig. 1. Basic ANC block diagram.

Methods to reduce excessive noise have two different categories: Passive noise control and active noise control. Passive noise control methods utilize sound absorbing materials such as glasswool. Sometimes the material is cut into special geometries to enhance their sound absorption capabilities. The overall noise isolation capability depends on a number of factors such as sound frequency, material type, its geometry, and its thickness. The sound absorption coefficient grows with sound frequency, which is a common attribute of all passive noise isolation materials. This becomes harder for the material to absorb sound when the wavelength of the sound exceeds the thickness of the substance. As a result, passive noise reduction devices operate effectively at higher frequencies but lose a lot of their efficiency at lower frequencies. When low frequency noise dominates the noise spectrum in many real-world circumstances, the passive noise reduction strategy becomes less effective.

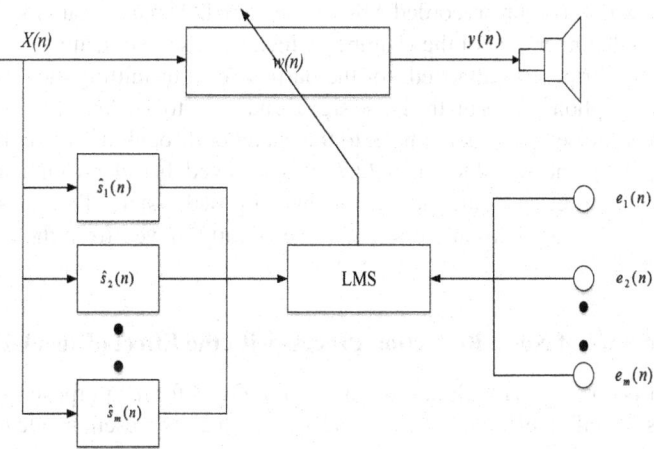

Fig. 2. Multi input single output FxLMS diagram.

Equation 5 describes the cost function $J(n)$ considering all M error sensors $e_m(n)$ and the secondary source input signal is given in Eq. 6. As illustrated in Fig. 2, each error sensor signal $e_m(n)$ is the addition of the primary noise source at the error microphone location $d_m(n)$ and the secondary noise source is formulated in Eq. 7.

$$J(n) = E\left\{\sum_{m=1}^{M} e_m^2(n)\right\} \tag{5}$$

$$y(n) = \sum_{i=0}^{N-1} w(i) \cdot x(n-i) \tag{6}$$

$$e_m(n) = d_m(n) + \sum_{i=0}^{l-1} \hat{s}_m(i) \cdot y(n-i) \tag{7}$$

where, $\hat{s}_m(i)$ denotes the estimation of the acoustic path between the loudspeaker and the mth error sensor.

Equation 4 developed the update step of the adaptive filter $w(n)$ in the same way. Equation 8 shows the weight update process.

$$w(n + 1) = w(n) + \mu \left(\sum_{m=1}^{M} e_m(n) \cdot \sum_{i=0}^{l-1} \hat{s}_m(i) \cdot x(n - i) \right) \qquad (8)$$

4   Experimental Setup

A glass chamber of 30 cm diameter and 2 mm thickness is employed. A MEMS microphone is attached at the top of the glass chamber to pick up the sound produced inside the chamber, which is a pre-recorded MRI noise signal. Three U sound speaker strips are attached to the inner side of the chamber, which acts as the transmitter of the source sound and three stripes are attached, for the purpose of transmitting the reduced noise signal. The microphone picks up the noise signal and gives to the ANC kit for processing and the reduced noise signal is fed back to the chamber through the secondary speaker stripes. A significant noise reduction of 22.7 dB is achieved. In order to enhance the noise reduction further, three cases are included in this proposed system. The measurement of sound level can be varied when the microphone slightly moved from the center of the tube.

4.1   Performance of Noise Reduction Process with the Effect of Insulation

The chamber is covered with glass wool, as seen in Fig. 3 for the purpose that no sound from the outside will interfere with the sound signal which is present inside the chamber and the sound signal which is inside will be reflected and the intensity will be increased. Noise reduction of 24 dB is achieved. The variation in SPL with glass wool insulation is depicted in Table 1 and Fig. 4. The SPL can be varied with the distance variation from microphone. Also the SPL is calculated with and without ANC. When the microphone is placed correctly at the center of the tube, the sound produced is 74 dB (without ANC). If the microphone moved away from the center, the sound level reduced. Similarly with ANC, the sound level is 72.6 dB, if the microphone at center. When it moved away, the sound level reduced.

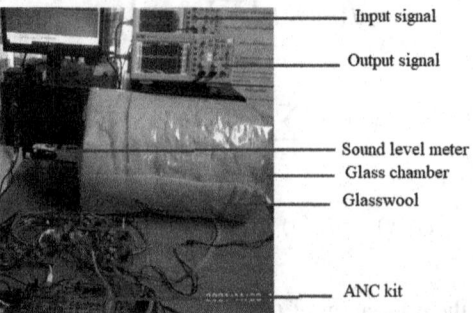

Fig. 3. Experimental set up with glass wool as insulation.

Table 1. Effect of insulation in ANC process.

Horizontal length of the microphone from the reference point (cm)	Vertical length of the microphone from the reference point (cm)	Sound Level when ANC is OFF (dB)	Sound Level when ANC is ON (dB
0	5	74	72.6
	5.5	76	68.3
	6	77.7	67.3
	6.5	85	80.3
1	5	74.9	75.3
	5.5	75.3	72.1
	6	77.8	69.2
	6.5	85	69.2
2	5	75.5	75
	5.5	73.7	72.2
	6	76.8	69.5
	6.5	83	74.6
3	5	75.2	75
	5.5	74.8	73.5
	6	76	74.3
	6.5	81.5	75.3
4	5	76	77.9
	5.5	75	77
	6	76.8	78.7
	6.5	80	78.4
5	5	77	77.8
	5.5	75	78.2
	6	75.8	78.8
	6.5	78.6	79.9
6	5	79.5	79.5
	5.5	80.2	73.5
	6	80.4	69.2
	6.5	80.5	68.5
7	5	81.6	63.2
	5.5	82.2	61.6
	6	83.4	58.3
	6.5	84.1	59.9

The performance is based on sound and distance, where the microphone position is varied horizontally and vertically and measurement is done when ANC circuit is OFF and ON. The FIR filter 64-order effectively implements both unwrap phase and magnitude

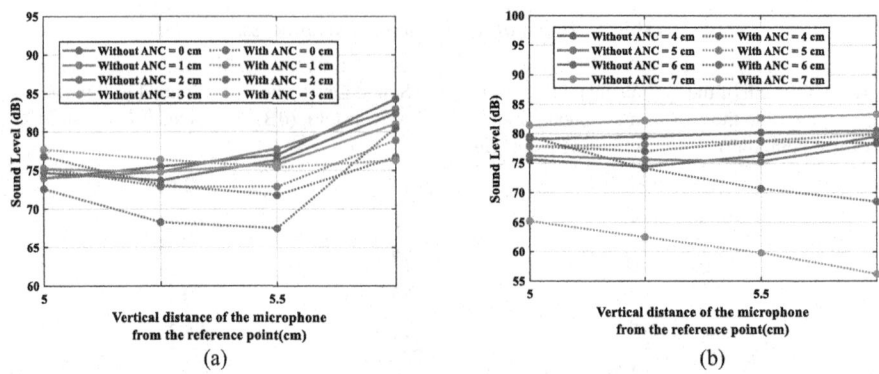

(a)   (b)

Fig. 4. Graphical representation of effect of insulation. (Color figure online)

response. The experiment is tested based on pre recorded MRI signal of frequency close to 4 kHz. The blue line represent the original noise when ANC is OFF and redline represent the output noise when ANC is applied. The insulation in proposed system when ANC is applied, the performance of broadband noise cancelling is still good and higher range is obtained.

4.2   Performance of Noise Reduction in the Presence of Static Magnetic Field

An array of magnets is attached to the inner surface of the chamber, as seen in Fig. 5 to create the magnetic field of 5000 Gauss, similar to that produced by the MRI machine and the sound reduction process is then analyzed. There is no effect in noise reduction because of the presence of magnetic field.

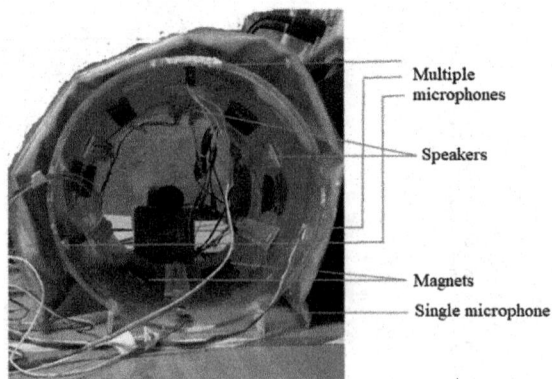

Fig. 5. Proposed hardware with magnets and multiple microphones

From the analysis of Table 2 and Fig. 6, the SPL does not depend on the magnetic field. There is no variation when there is a magnetic field present in the system. Horizontal distance from reference point are varied from 0 cm to 7 cm and the results are evaluated.

Table 2. Influence of magnetic field in ANC process.

Horizontal length of the microphone from the reference point (cm)	Vertical length of the microphone from the reference point (cm)	Sound Level when ANC is OFF (dB)	Sound Level when ANC is ON (dB
0	5	74	72.6
	5.5	76	68.3
	6	77.7	67.3
	6.5	85	80.3
1	5	74.9	75.3
	5.5	75.3	72.1
	6	77.8	69.2
	6.5	85	69.2
2	5	75.5	75
	5.5	73.7	72.2
	6	76.8	69.5
	6.5	83	74.6
3	5	75.2	75
	5.5	74.8	73.5
	6	76	74.3
	6.5	81.5	75.3
4	5	76	77.9
	5.5	75	77
	6	76.8	78.7
	6.5	80	78.4
5	5	77	77.8
	5.5	75	78.2
	6	75.8	78.8
	6.5	78.6	79.9
6	5	79.5	79.5
	5.5	80.2	73.5
	6	80.4	69.2
	6.5	80.5	68.5
7	5	81.6	63.2
	5.5	82.2	61.6
	6	83.4	58.3
	6.5	84.1	59.9

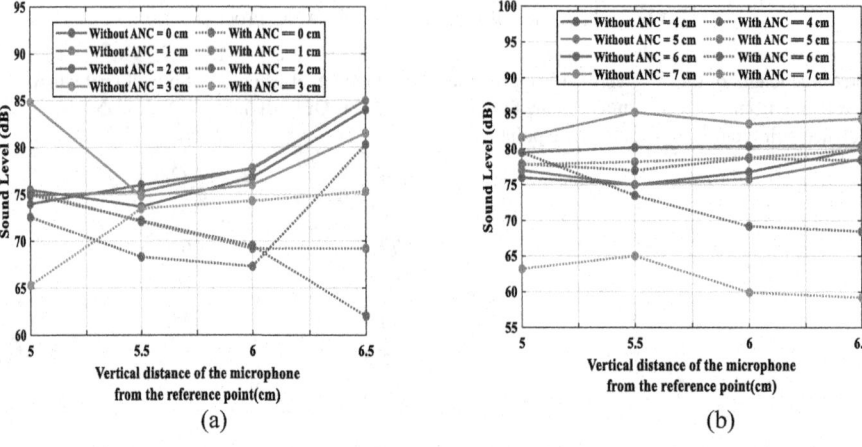

Fig. 6. Graphical representation of effect of magnetic field for with and without ANC. (Color figure online)

4.3 Performance of Noise Reduction with Multiple Microphones

MEMS microphones are used within the chamber to pick up the sound signal and multiplexed. MEMS microphones sensitive range lies between −46 dBV and −35 dBV. This level is a better concession between the noise floor. The maximum acoustic input is normally about 120 dB SPL. The output from digital MEMS microphones and low output from analog MEMS microphones are perfect application for electrically noise source. MEMS microphone has omnidirectional directivity that collect a sound equally from several direction. The resultant signal is given to the ANC kit for further processing. The basic arrangement of a single microphone is used at the center of the chamber to pick up the sound signal and given to ANC kit. Additionally, three microphones are attached at the side of the glass chamber at proper spacing to pick up the noise. The sound signals are processed in the ANC kit, where it is reduced and given back to the chamber. Noise reduction of 25 dB is achieved when multiple microphones are used. This is greater than noise reduction achieved with single microphone.

From the Table 3, and Fig. 7 analysis, the SPL is varied with the multitude microphones. The noise reduction of 25 dB is achieved.

The performance is based on sound and distance. It is effective and simple ANC and improves the patient comfort during scanning. The noise reducing zone is generated to improve the quality. The experiment is also tested in real time with 1.5 Tesla MRI machine and attained good result.

Table 3. ANC variation with effect of multiple microphones.

Horizontal length of the microphone from the reference point (cm)	Vertical length of the microphone from the reference point (cm)	Sound Level when ANC is OFF (dB)	Sound Level when ANC is ON (Db
0	5	74	72.6
	5.5	76	68.3
	6	77.7	67.3
	6.5	85	80.3
1	5	74.9	75.3
	5.5	75.3	72.1
	6	77.8	69.2
	6.5	85	69.2
2	5	75.5	75
	5.5	73.7	72.2
	6	76.8	69.5
	6.5	84	62
3	5	84.8	65.3
	5.5	74.8	73.5
	6	76	74.3
	6.5	81.5	75.3
4	5	76	77.9
	5.5	75	77
	6	76.8	78.7
	6.5	80	78.4
5	5	77	77.8
	5.5	75	78.2
	6	75.8	78.8
	6.5	78.6	79.9
6	5	79.5	79.5
	5.5	80.2	73.5
	6	80.4	69.2
	6.5	80.5	68.5
7	5	81.6	63.2
	5.5	85.1	65
	6	83.5	59.9
	6.5	84.2	59.2

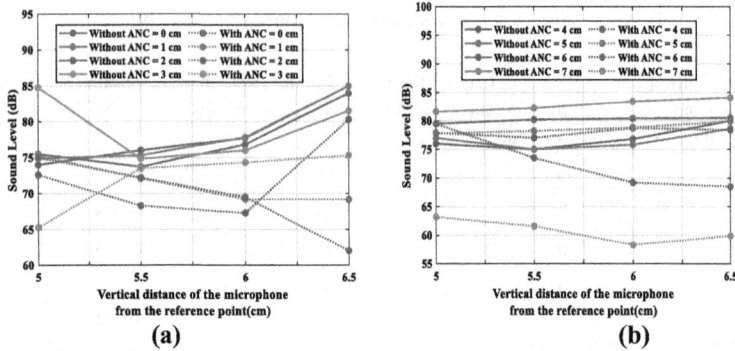

Fig. 7. Graphical representation of Noise reduction with multiple microphones. (Color figure online)

5 Conclusions

An ANC system was successfully developed and tested on a MRI system used for clinical imaging. The ANC system reduces the overall noise level at the patient by as much as 22.7 dB. Then the performance of the proposed system can be measured using three cases such as insulation, magnetic field, and multiple microphones. From the analysis, the performance is varied with these cases. By using the glass wool insulation, there is a noise reduction of 24 dB, while the performance of the system does not depend on the magnetic field. The multiple microphone usage helps in achieving a noise reduction of 25 dB. The system is examined with real time MRI signal. The outcomes shows that the total noise level heard by the patient is less than the noise level in MRI system without ANC. The frequency range can be optimized as a future scope of this study along with study on application of various algorithms in the frequency domain for optimization.

References

Ardekani, I.T., Abdulla, W.H.: Adaptive signal processing algorithms for creating spatial zones of quiet. Digit. Sig. Process. **27**, 129–139 (2014)

Chang, C.Y., Pan, S.T., Liao, K.C.: Active noise control and its application to snore noise cancellation. Asian J. Control **15**(6), 1648–1654 (2013)

Furlong, T.S., Anderson, B.E., Patchett, B.D., Sommerfeldt, S.D.: Active noise control using remotely placed sources: application to magnetic resonance imaging noise and equivalence to the time reversal inverse filter. Appl. Acoust. **176**, 107902 (2021)

Hasegawa, R., Yamashita, H., Kajikawa, Y.: Effectiveness of active noise control system for nonstationary noise in consideration of psychoacoustic properties. In: 2018 Asia-Pacific Signal and Information Processing Association Annual Summit and Conference (APSIPA ASC) , pp. 1256–1261. IEEE (2018)

Hennel, F.: Fast spin echo and fast gradient echo MRI with low acoustic noise. J. Magn. Reson. Imaging Off. J. Int. Soc. Magn. Reson. Med. **13**(6), 960–966 (2001)

Huang, C.-R., Chang, C.-Y., Kuo, S.M.: Implementation of feedforward active noise control techniques for headphones. In: 2020 Asia-Pacific Signal and Information Processing Association Annual Summit and Conference (APSIPA ASC) , pp. 293–296. IEEE (2020)

Hussain, Z., Gimenez, F., Yi, D. Rubin, D.: Differential data augmentation techniques for medical imaging classificationtasks. In: AMIA Annual Symposium Proceedings, vol. 2017, p. 979. American Medical Informatics Association (2017)

Kang, L., Xia, L.: Acoustic control through gradient coil design using a finite-difference-based method for MRI. Discrete Dyn. Nat. Soc. **2021** (2021)

Kumamoto, M., Kajikawa, Y., Tani, T., Kurumi, Y.: Active noise control system for MR noise. In: Proceedings of the Second APSIPA Annual Summit and Conference, pp. 543–548 (2010)

Kuo, S.M., Chen, Y.R., Chang, C.Y., Lai, C.W.: Development and evaluation of light-weight active noise cancellation earphones. Appl. Sci. **8**(7), 1178 (2018)

Lee, N., Park, Y., Lee, G.W.: Frequency-domain active noise control for magnetic resonance imaging acoustic noise. Appl. Acoust. **118**(2017), 30–38 (2017)

Li, X., Qiu, X., Leclercq, D.L.L., Zander, A.C., Hansen, C.H.: Implementation of active noise control in a multi-modal spray dryer exhaust stack. Appl. Acoust. **67**(1), 28–48 (2006)

Lu, L., et al.: A survey on active noise control in the past decade–part I: linear systems. Sig. Process. 108039 (2021)

Luo, L., Sun, J., Huang, B.: A novel feedback active noise control for broadband chaotic noise and random noise. Appl. Acoust. **116**, 229–237 (2017)

McJury, M., Stewart, R.W., Crawford, D., Toma, E.: The use of active noise control (ANC) to reduce acoustic noise generated during MRI scanning: some initial results. Magn. Reson. Imaging **15**(3), 319–322 (1997)

Meng, H., Chen, S.: A modified adaptive weight-constrained FxLMS algorithm for feedforward active noise control systems. Appl. Acoust. **164**, 107227 (2020)

Muto, K., Nakayama, S., Osada, R., Yagi, K., Chen, G.: A study of the position of the reference microphone of active noise control of feedforward type for MRI noise. In: INTER-NOISE and NOISE-CON Congress and Conference Proceedings. Institute of Noise Control Engineering, vol. 249, no. 8, pp. 282–287 (2014)

Pla, F.G.: Active control of noise in magnetic resonance imaging. Active **95**, 573–583 (1995)

Rondinoni, C., Amaro, E., Cendes, F., Santos, A.C.D., Salmon, C.E.G.: Effect of scanner acoustic background noise on strict resting-state fMRI. Braz. J. Med. Biol. Res. **46**, 359–367 (2013)

Roozen, N.B., Koevoets, A.H., Hamer, A.J.D.: Active vibration control of gradient coils to reduce acoustic noise of MRI systems. IEEE/ASME Trans. Mechatron. **13**(3), 325–334 (2008)

Rudd, B.W., Lim, T.C., Li, M. Lee, J.H.: In situ active noise cancellation applied to magnetic resonance imaging (2012)

Sartoretti, E., et al.: Impact of acoustic noise reduction on patient experience in routine clinical magnetic resonance imaging. Acad. Radiol. (2020)

Yamashiro, T., Morita, K., Nakajima, K.: Evaluation of magnetic resonance imaging acoustic noise reduction technology by magnetic gradient waveform control. Magn. Reson. Imaging **63**, 170–177 (2019)

Rough Rice Grading in the Philippines Using Infrared Thermography

Orfel L. Bejarin[1][(✉)] and Arnel C. Fajardo[2]

[1] College of Engineering Architecture and Technology, Isabela State University, 3300 Ilagan, Isabela, Philippines
orfel.l.bejarin@isu.edu.ph
[2] College of Computing Studies, Information and Communication Technology, Isabela State University, 3305 Cauayan City, Isabela, Philippines

Abstract. The Rough rice grading is a set of standard methods and processes for evaluating quality that are important to marketing and process controls. IR thermography uses the electromagnetic spectrum's infrared region to detect heat that has been emitted. FLIR ONE thermal camera was used to acquire images of the sample. In order to represent each sample, a thermal index was generated using the average value of the pixels in the thermal image. In this study, the grading of rough rice was evaluated using infrared thermography and image processing techniques. Using infrared thermography, the results of the study indicated that detecting moisture content was 87.30% accurate and detecting foreign matter was 95.04% accurate. The results showed that infrared thermography evaluation using spectral analysis could accurately determine the moisture content of rough rice. Farmers, rice millers, and other related government agencies can greatly benefit from the application of non-contact, non-destructive methods for determining the purity and moisture content of rough rice.

Keywords: Infrared Thermography · Moisture Content · CNN · Image Processing · Rough Rice · Confusion Matrix · FLIR Thermal Studio · Matlab

1 Introduction

The Philippines has long been renowned as an agricultural country and one of its agricultural crops is rice. The majority of Filipinos rely on agriculture to support their daily necessities. They are particularly interested in farming because it allows them to obtain food while also earning money.

In Philippines, the rough rice has been graded based on the moisture content, purity and foreign matters of rough rice, based from Philippine National Standard PNS/BAFS 290:2019 Grains – Grading and classification – Rough rice and milled rice. It was used by National Food Authority (NFA) and rice millers to give price to the farmer when purchasing rough rice. The rough rice is classified as extra-long, long, medium and short. The maximum moisture content for rough rice shall be in conformance with the set value which is 14% and should conform with the grade requirements specified by

F. Neri et al. (Eds.): CCCE 2023, CCIS 1823, pp. 16–26, 2023.
https://doi.org/10.1007/978-3-031-35299-7_2

the PNS which are Premium Grade, Grade no. 1, Grade no. 2, and Grade no. 3. Based from the purity and foreign matter of the rough rice [1].

Precision agriculture with the use of ICT in specific farm areas will bring about a new level of technology to reduce costs, improve efficiency and provide insights for improvement. While the extent of use is yet very limited, but this is churning interests especially for the young farmers and make them realize that rice agriculture is not just dirty work, but can be made smarter with the use of advance technology [2]. Rice quality can be classified based on the form, color and moisture content of the rice grain [3].

Research indicates that rice grading is a process of sorting rice and assign into its classes or grade. The grading of rice plays important role in the determination of rice quality method applied in the rice production industry and its subsequent price in the market. The grade and price of rice is largely determined by its quality, genetic, agronomic and commercial value [4]. Rice grading can be done in varieties of ways based on its special features. For examples like color approach using filter, edge detection template, decision tree analysis and many more. Depending on the goal of classification that the user wants to obtain. For example, for the goal such as originality of rice, color and length features can be used. Machine vision system development has been capable and helpful in sorting rice into chalky, cracked, broken and internal damaged kernels. These are some approaches which can be used and helpful in rice classification and grading [5].

In the agricultural sector, it is crucial to quantify the moisture content (MC) of rough rice. Application of a non-destructive approach to measure MC is required due to rising rice product prices. Additionally, by measuring one rice kernel, we may rapidly determine the MC distribution of the rice samples. Rice is often harvested when it has a high moisture content (MC), typically between 25 and 40% dry basis (db) or 25 and 32% wet basis (wb). At this moisture level, rice has a high respiration rate and is particularly vulnerable to assault from microorganisms, insects, and pests. The average MC of the grain bulk is important in marketing. As an instance, 1,500 tons of water are represented by a cargo of 10,000 tons with 15% (w.b. is used herein unless otherwise specified) MC. That is worth USD 10,000 at a cost of USD 0.10 per kg. The quantity of this pricey water that is involved must be revealed to both the buyer and the seller [6].

Moisture content (MC) has significant impact on many mechanical properties of wood. Oven-dry basis MC is commonly used in lumber industry and is defined as the amount of water-mass existing in the material divided by oven-dry mass. MC is mostly represented as a percentage of oven-dry mass. However, Oven-dry method is not the only way of measuring MC. We can estimate MC of specimen using relative humidity and temperature of the environment. If given enough time, wood specimens will eventually reach a moisture equilibrium with the environment depending on the temperature and relative humidity. The MC at this point is called Equilibrium moisture content (EMC) and can be estimated using relative humidity and temperature [7].

Another researcher mentioned in their research that Pearson's correlation coefficient has been proven to identify the relationship between thermal index to the moisture content of rough rice. The researcher also emphasizes that Pearson's correlation coefficient assumes that each pair of variables has a bivariate normal distribution. The closer the value of correlation to 1, the closer the two variables to a perfect positive correlation, while the closer the value to -1, the closer the two variables to a perfect negative correlation [8].

However, some research pointed that quality evaluation and assessment is done by human sensory panel which is time consuming and there is deviation in results and expensive. This can be replaced with Computer vision and Neural network [9]. For quality analysis, maximum numbers of parameters are to be measured by image processing techniques. Expansion on this work can target to design such a system which can classify rice grains on the basis of each parameter which can used to enhance the quality of rice. The cost of such system should be less and minimize time requirement for quality analysis [10]. Focusing on different sampling methods, sample sizes, sample pre-processing techniques, different features and different neural network models to match the requirements of the rice industry [11]. Color and texture features can be extracted to increase the accuracy rate. Also, detection of various defects of rice kernels, like fissures, can be investigated [12]. Performing fast, non-destructive and accurate MC measurements is crucial in agriculture, industry, storing, trading, and processing of rice products. If instrument can offer both mean and variation of MC samples, it would be more helpful for processing and pricing [6].

The aforementioned studies have established the necessity for an effective method of grading rough rice in the Philippines that benefits the government, farmers, rice millers, and consumers. Both the buyer and the seller might benefit from higher sales by using infrared thermography to grade rough rice. The goal of the current study is to achieve this. The aforementioned studies have established the necessity for an effective method of grading rough rice in the Philippines that benefits the government, farmers, rice millers, and consumers. Both the buyer and the seller might benefit from higher sales by using infrared thermography to grade rough rice. The goal of the current study is to achieve this.

2 Materials and Methods

2.1 Rough Rice Sample

The samples of rough rice used in this study were taken from a rice granary in the Philippines. To get samples of actual rough rice, the samples were also taken during the harvest season and before drying. Three (3) various rough rice cultivars that are frequently grown in the Philippines were used in this study.

Rough rice is typically harvested around 20% to 25% wet basis moisture. A Smart Sensor AR991 Grain Moisture Meter was used to calculate the rough rice's moisture content. Rough rice with moisture content ranging from 11% to 55% may be examined with a grain moisture meter. Rough rice was measured three times, and the average moisture content was found. The data was used to demonstrate the accuracy in determining the moisture content of the rough rice.

Using a sun drying technique similar to what Filipino farmers do, the samples' actual moisture content was measured. In order to determine the rough rice's purity and its moisture content, samples of the rough rice were prepared for each of these two purposes. There were 2,238 samples analyzed in this study for grading and classification.

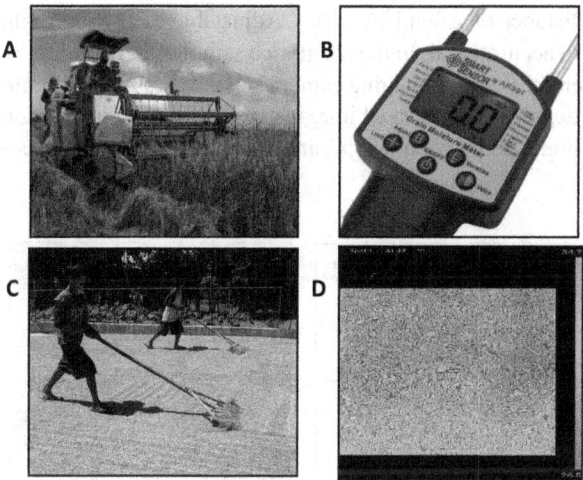

Fig. 1. (A) Rough rice harvesting; (B) AR991 grain moisture meter; (C) Sun drying; (D) IR Image of rough rice.

The procedure for collecting samples, from harvesting to drying to classification, is shown in Fig. 1.

2.2 Image Acquisition

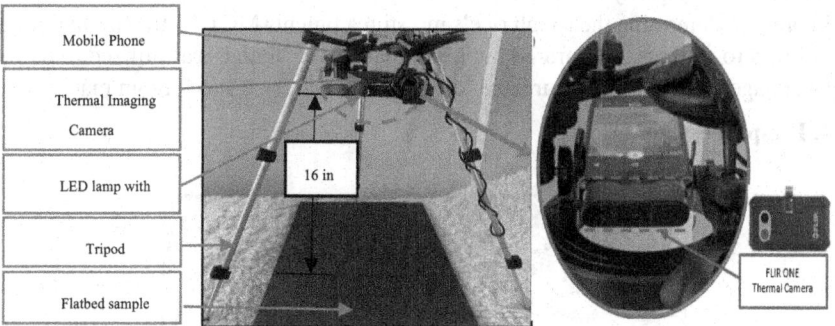

Fig. 2. Thermal imaging setup.

In this Figure, IR Images of rough rice samples was captured using a thermal imaging camera (FLIR ONE Gen 3) with an 80×60 thermal resolution and a visual resolution of 1440×1080, which were attached to a mobile phone device and saved in FLIR ignite cloud storage. FLIR ONE utilizes Multi Spectral Dynamic Technology (MSX) to combine the regular and thermal images [13]. This will be done to provide the raw thermal image additional raw data, resulting in a better vision as RGB scaled images. At least 1500 images for each rough rice sample were captured. The samples are obtained with a sampling distance of around 16 inches. Adjustable LED Ring Light (5 Watts to 75 Watts) was used to accurately light the rough rice samples and eliminate harsh shadows. The angle between the thermal imaging camera and the axis of the illumination source is kept at about 45 degrees. The thermal imaging camera is mounted on a modified tripod as illustrated in Fig. 2 to provide a rigid and stable support as well as simple vertical mobility.

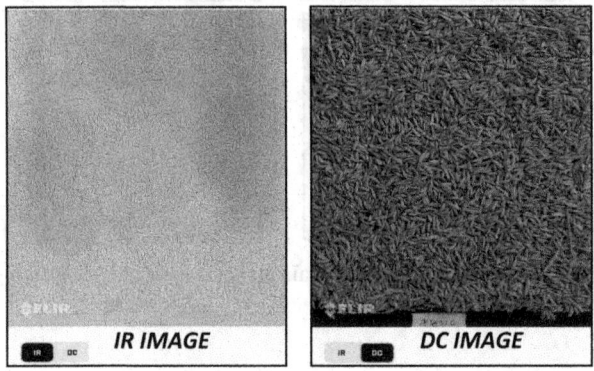

Fig. 3. Infrared image acquired from FLIR ONE thermal camera.

Figure 3 shows the acquired Infrared image from the FLIR ONE thermal camera, this device can generate two types of images with single shot: an Infrared (IR) image and a Digital Color (DC) image. The FLIR ONE camera's unique features may be utilized to powerful advantage by researchers.

In order to determine the rough rice's moisture content (MC), FLIR Thermal Studio was utilized to process the infrared images. Meanwhile, Matlab was utilized to process the DC image in order to measure and identify the rough rice and foreign matter.

2.2.1. Pre-processing

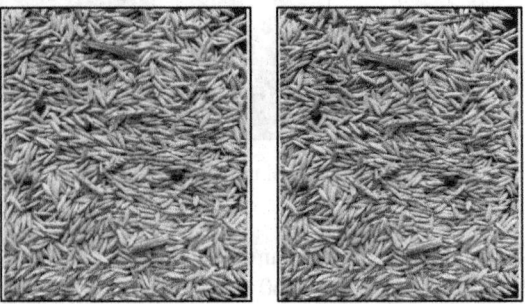

Fig. 4. Converted DC image to grayscale image.

This figure shows the conversion of an infrared image from a DC to a grayscale representation. The noise and background from the DC image were removed by filtering.

Noise in raw photos needs to be removed frequently. It includes smoothing, offset, and normalization. The infrared images should then be converted into grayscale images (Fig. 4).

2.2.2. Segmentation is the process of partitioning an IR image up into several elements. This is frequently used to identify the object and other significant details in an IR image. The process of segmenting and extracting the region of interest (ROI) from a grayscale image is exceedingly difficult. The grayscale IR image is segmented into several segments as a group of pixels or super-pixels in order to detect objects in the grayscale images. A binary image may be generated from a greyscale image using a threshold.

Finding the ideal threshold to distinguish the foreground from the background is necessary for the histogram shape thresholding. Dark and bright areas are two regions that show the presence of two separate regions. Using semantic image segmentation, the optimum threshold was calculated. Labels for the background and foreground pixels were assigned using the calculated threshold.

Semantic Segmentation is an important part of image object detection task. It is the understanding of an image at pixel level. Most methods for semantic segmentation require each pixel with an object label in the image. It makes the prediction at every pixel. The prediction is about not only the class but also the boundaries of each kind object [14].

2.2.3. Edge Detection

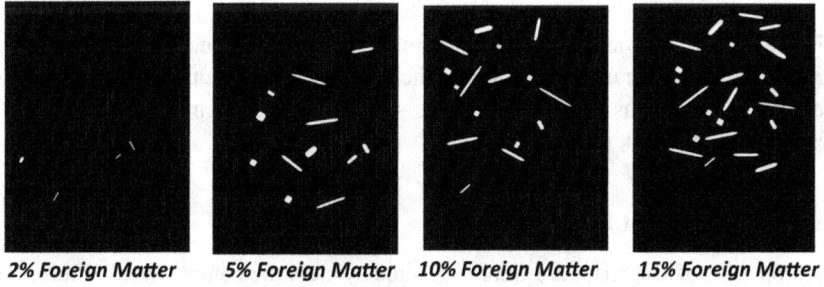

2% Foreign Matter   5% Foreign Matter   10% Foreign Matter   15% Foreign Matter

Fig. 5. Segmented foreign matter from the binary image.

Figure 5 shows the segmented image of foreign matter from the binary image with the corresponding percentage of foreign matter. Edge detection is the groundwork for edge detection using various edge operators. Edge operators can identify variations in texture, color, and other attributes. Canny edge detection is the best detector for producing the best filtered image. This ideal detecting method finds the borders of the grey scale image.

2.2.4. Features Extraction is the identification of the binary image's significant qualities and measurable properties is known as features extraction. The identification of various regions in an image can be aided by the image's extracted features. Both low-level and high-level characteristics are possible for these features. The image intensity values are used to directly extract the low-level features.

2.3 Purity and Foreign Matter

The amount of rough rice that is free of foreign matter is known as purity. Foreign matter, on the other hand, is any substance that is not rough rice, including chaff, straw, other crop seeds, gravel, soil, stones, clay, and mud. The frequency and percentage distribution method will be used in this study to measure the rough rice's purity, and the foreign matter were converted into binary images using segmentation and thresholding and were processed in MATLAB.

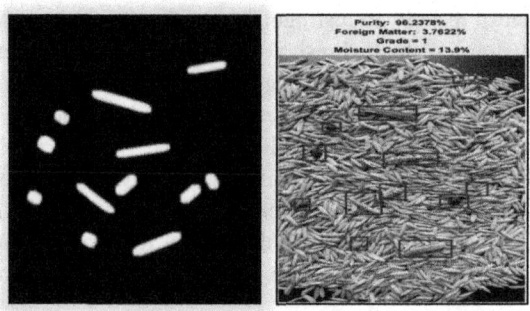

Fig. 6. Detection and computation of total pixel of the foreign matter and rough rice to determine the purity of the rough rice.

Figure 6 shows the detection of foreign matter and rough rice. The system will first detect the foreign matter using Matlab, and then compute for the total pixel of the foreign matter (White pixels), as seen in Fig. 6's left-hand corner. Rough rice quantity will be represented by the black pixel area.

2.4 Moisture Content

Many properties of rough rice are greatly affected by moisture content (MC). The amount of water in the rough rice divided by the oven-dry mass is known as oven-dry basis MC, and it is widely employed in the agriculture sector. But there are other ways to measure MC besides the oven-dry method.

Using the extracted information from a IR image of the rough rice, we can measure the MC of the sample using the relative humidity and temperature. Rough rice sample will ultimately achieve moisture equilibrium with the experiment based on the temperature and relative humidity. Using FLIR Thermal Studio Software, temperature and relative humidity were measured.

Equilibrium Moisture Content (EMC), can be calculated using the equation below. The relative humidity (h) is expressed in decimals in the equation below (e.g., 0.70 instead of 70%), and T is the temperature (in degrees Fahrenheit).

$$W = 330 + 0.452T + 0.00415T^2 \tag{1}$$

$$K = 0.791 + 0.000463T - 0.000000844T^2 \tag{2}$$

$$K_1 = 6.34 + 0.000775T - 0.0000935T^2 \tag{3}$$

$$K_2 = 1.09 + 0.084T - 0.0000904T^2 \tag{4}$$

where T is the dry-bulb temperature (°F). Thus, given two pieces of information, dry-bulb (or ambient) temperature and the RH, the EMC can be readily calculated.

$$EMC(\%) = \frac{1800}{W}\left[\frac{Kh}{1 - Kh} + \frac{K_1 Kh + 2K_1 K_2 K^2 h^2}{1 + K_1 Kh + K_1 K_2 K^2 h^2}\right] \tag{5}$$

where EMC is the equilibrium moisture content (%), h is the relative humidity expressed in decimal form (%/100), and $W, K, K1$, and $K2$ are coefficients defined by Eqs. 1 through 4, respectively [15]

2.5   Evaluation Using Confusion Matrix

Evaluation is the last part of the process after determining the moisture content and the purity of the rough rice. A confusion matrix is a table that is often used to describe the performance of a classification model (or "classifier") on a set of test data for which the true values are known. It allows the visualization of the performance of an algorithm. It allows easy identification of confusion between classes [16]. The *Confusionmat* function, one of the useful Matlab functions, was utilized by the researcher [17, 18]. The accuracy of the model can be easily calculated using the confusion matrix. The formula for calculating accuracy is:

$$Accuracy = \frac{TP + TN}{TP + TN + FP + FN} \tag{6}$$

3   Results and Discussion

Foreign Matter
The confusion matrix obtained by training a classifier and evaluating the trained model on this test set is shown below.

CONFUSION MATRIX

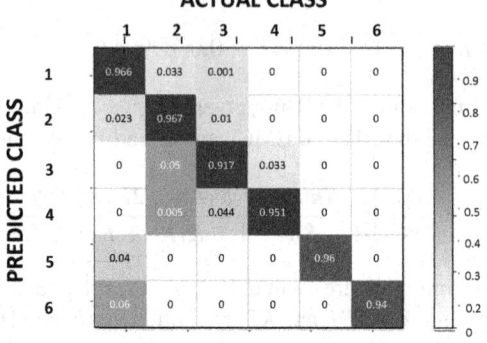

Fig. 7. The Confusion matrix for the determination of foreign matters and rough rice with accuracy rate of 95.04%.

The classification confusion matrix generated by the Discriminant Analysis method is shown in Fig. 7 for each class. In Fig. 7, there are three foreign matter classes that obtained high accuracy rates: class 2 or (2% foreign matter) at 96.70%; class 1 with (5% foreign matter); and class 5 with (100% foreign matter), which had accuracy rates of 96.60% and 96.00%, respectively. Class 4 came in second with (15% foreign matter), while class 6 came in third with (100% rough rice) with accuracy percentages of 95.10% and 94.00%, respectively. On the other hand, the Class 3 with 10% foreign matter had the lowest accuracy rate for classification. The overall classification accuracy rate is, in fact, 95.04%. The lighting situation was cited for confusion and misclassifications [19].

Moisture Content

CONFUSION MATRIX

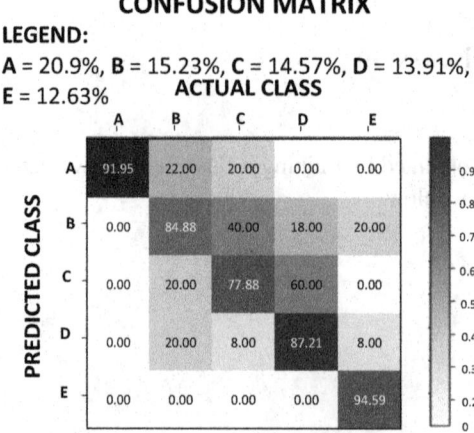

Fig. 8. The confusion matrix for moisture content detection with accuracy rate of 87.30%.

Figure 8 shows that the diagonal components are the predicted observations. Out of the 2,238 observations, a total of 2,2002 were correctly predicted. The total accuracy is therefore 87.30%. Figure 8 indicates that Class E - 12.63%MC had the highest accuracy rate for classification (94.59%), followed by Class A-20.29%MC (91.95%), then Class D-14.57 and Class B-17.52 (87.21% and 84.88%, respectively). Nevertheless, Class C-15.23 seemed to have the lowest accuracy rate for classification, at 77.88%. Focusing on the Class C-15.23 predicted outcomes will contribute to improving the model's performance. The classifier incorrectly classified 80 samples collectively from Class C-15.23—the highest misclassification rate of any class—with the possible exception of the sample. For Class C-15.23, this implies that 77.88% of predictions were accurate.

In one study, the efficiency of impedance and near-infrared spectroscopy for determining out the moisture content of rough rice was evaluated. The results showed that both techniques were extremely accurate, with near-infrared spectroscopy having a marginally higher accuracy (0.55% error) than impedance (0.79% error) [20, 21].

4 Conclusion

Based on the results of this study, it can be concluded that the purity of rough rice may be determined using an image processing algorithm that determines the average pixel of foreign matter and rough rice. By analyzing the presence of relative humidity and temperature from the IR image and process using FLIR thermal studio, the moisture content measurement can be scientifically determined using the equilibrium moisture content to predict the moisture content of the rough rice.

The temperature difference between foreign objects and rough rice may be observed through hyperspectral. The average success rate for using the thermal imaging algorithm to obtain foreign matter is 95.04%, while the accuracy rate for detecting moisture content is 87.30%. The results of this study indicate that the detection of foreign matter and rough rice's moisture content could be achievable using infrared thermography technique. However, research on the implementation of this technology can be done in the future, and it is strongly suggested to develop new techniques for image enhancement, feature extraction, machine learning, dataset creation, and real-time rough rice classification system that can classify rice in real-time during production can help ensure that the quality of rice meets the required standards. Researchers can improve the accuracy and efficiency of rice classification, which is critical for ensuring food security and meeting the growing demand for high-quality rice.

References

1. Bureau of Agriculture and Fisheries Standards: PHILIPPINE NATIONAL STANDARD Grains – Grading and Classification - Corn. (632) (2017)
2. Tallada, J.G.: Precision Agriculture for Rice Production in the Philippines. FFTC Agricultural Policy Platform (2019)
3. Prihasty, W., Nasution, A.M.T., Isnaeni: Rice grain quality determination using FTIR spectroscopy method. In: 2020 IEEE 8th International Conference on Photonics (ICP) (2020). https://doi.org/10.1109/icp46580.2020.9206464

4. Gudipalli, A., , N, A.P.: A review on analysis and grading of rice using image processing. ARPN J. Eng. Appl. Sci. **11**(23), 13550–13555 (2016)
5. Tahir, W.P.N.W.M., Hussin, N., Htike, Z.Z., Yan, W., Naing, N.: Rice grading using image processing. ARPN J. Eng. Appl. Sci. **10**(21), 10131–10137 (2015)
6. Heman, A., Hsieh, C.-L.: Measurement of moisture content for rough rice by visible and near-infrared (NIR) spectroscopy, engineering in agriculture. Environ. Food (2016). https://doi.org/10.1016/j.eaef.2016.02.002
7. TIMBERAID.com, Wood Moisture Content. https://www.timberaid.com/calculator/fundamental/moisturecontent. Accessed 11 May 2020
8. Bejo, S.K., Jamil, N.: Paddy grading using thermal imaging technology. Int. Food Res. J. **23**, 245–248 (2016)
9. Jain, N.K., Khanna, S.O., Jain, K.R.: Development of a classification system for quality evaluation of Oryza Sativa L. (Rice) using computer vision. In: Proceedings - 2014 4th International Conference on Communication Systems and Network Technologies, CSNT 2014, pp. 1088–1092 (2014). https://doi.org/10.1109/CSNT.2014.222
10. Mahale, B., Korde, P.S.: Rice quality analysis using image processing techniques. In: International Conference for Convergence of Technology, pp. 1–5 (2014)
11. Silva, C.S.,Sonnadara, U.: Classification of rice grains using neural networks. In: Proceedings of Technical Sessions, Sri Lanka, September 2013, pp 9–14 (2013)
12. Mousavirad, S.J., Akhlaghian Tab, F., Mollazade, K.: Design of an expert system for rice kernel identification using optimal morphological features and back propagation neural network. Int. J. Appl. Inf. Syst. **3**, 33–37 (2012)
13. TELEDYNE FLIR. What is MSX? https://www.flir.asia/discover/professional-tools/what-is-msx/. Accessed 05 June 2020
14. Li, B., Shi, Y., Qi, Z., Chen, Z.: A Survey on Semantic Segmentation, pp. 1233–1240 (2018). https://doi.org/10.1109/ICDMW.2018.00176
15. Mitchell, P.: Calculating the equilibrium moisture content for wood based on humidity measurements. BioRes. **13**(1), 171–175 (2018). https://doi.org/10.15376/biores.13.1.171-175
16. knowledgebank.irri.org. How to determine the EMC. http://www.knowledgebank.irri.org/step-by-step-production/postharvest/drying/drying-basics/how-to-determine-the-emc. Accessed 15 Mar 2020
17. Arora, V.: Confusion matrix. https://rstudio-pubs-static.s3.amazonaws.com/559365_9d524b0feb814087a2b4db7ba41df323.html. Accessed 15 Mar 2020.
18. Panneton, B.: Gray image thresholding using the Triangle Method. https://www.mathworks.com/matlabcentral/fileexchange/28047-gray-image-thresholding-using-the-triangle-method. Accessed 07 Feb 2022
19. Marcelo, R., Lagarteja, J.: Corzea: Portable Maize (Zea Mays L.) Nutrient Deficiency Identifier (2021)
20. Muthukaruppan, K., et al.: A comparison of South East Asian face emotion classification based on optimized ellipse data using clustering technique. J. Image Graph. **3**(1), 1–5 (2015). https://doi.org/10.18178/joig.3.1.1-5
21. Zhou, Y., Chen, H., Zhang, X., Li, Z., Zhang, Y.: Comparative study on determination of moisture content in rough rice using near infrared spectroscopy and impedance. Food Anal. Methods **13**(6), 1416–1424 (2020)

An Interpretable Hybrid Recommender Based on Graph Convolution to Address Serendipity

Ananya Uppal[(⊠)], P. Maitreyi, P. Shreya, Trisha Jain, and Bhaskaryoti Das

PES University, Bengaluru, India
2712ananya@gmail.com
https://pes.edu

Abstract. This paper proposes a hybridized recommender system built to over-come the disadvantages of its individual components by combining them in a way that balances contrasting metrics. The individual com-ponents are a content-based and collaborative filtering-based model, a neural network model, and a graph convolutional network model. The individual models used in today's scientific landscape focus entirely on a single metric to be tuned and optimized. Through this paper, we introduce a way to balance all metrics, while retaining excellent precision and recall and improving on less focused metrics such as coverage and serendipity. We also explore a novel hybridization technique that represents the recommendation scenario as a graph and infers edges from the dataset to map relevant relationships. Interpretability is the comprehension of a model's funda-mental decision-making. This helps improve users' trust in the model and is an attempt at understanding the features and their relevance. The results of the hybrid recommender are explained using post hoc interpretability techniques. Addition-ally, serendipity is used to capture user satisfaction and the factor of "pleasant surprise" with the recommendations. To overcome the subjective nature of eval-uating serendipity, we also propose a new distance-based method to calculate it. The results provide a comparison of contrasting and competing metrics of the individual and hybrid models, and also show the balance of metrics achieved by the hybrid models.

Keywords: Recommender System · Neural Network · Graph Convolutional Network · Interpretability · Serendipity

1 Introduction

Recommender Systems are defined as filtering systems that aim to recommend an Item to a User by predicting the rating or preference the user would give to the item. With the terabytes of information available online today, recommender systems are crucial and necessary in preventing an overload of information and hence eliminating user over-choice. One of the earliest models for Recommender Systems is Matrix Factorisa-tion [1] and recently, Deep Learning techniques have been increasingly applied in the field of recommendation due to their ability to overcome the limitations of conventional models and provide high-quality recommendations [2]. The need to effectively capture

© The Author(s), under exclusive license to Springer Nature Switzerland AG 2023
F. Neri et al. (Eds.): CCCE 2023, CCIS 1823, pp. 27–38, 2023.
https://doi.org/10.1007/978-3-031-35299-7_3

non-Euclidean data has led to an increase in the use of Graph Neural Networks for recommendations. Graph Neural Networks effectively capture multi-hop relationships along with explicit and implicit relations between the users and items [3]. Each new model of recommendation, while providing considerable advantages to the recommendation problem, also come with their own set of limitations. This paper proposes a Hybrid Recommender System to overcome certain limitations along with a Novelty factor to evaluate the recommender's results. It also explains the hybrid model's working through interpretability.

The main contributions of this paper are as follows:

- A hybrid recommender system that balances the metrics of its individual components and benefits from their advantages.
- A new metric to evaluate serendipity of recommendations.
- An Interpretability technique to understand the results obtained from the hybrid model.
- A Graph-based recommender system that combines content-based, collaborative-filtering-based, and social information to provide recommendations.

The following sections describe the current landscape in the field of recommender systems, the methodology used by us to develop our model and to calculate the serendipity shown by the model, and finally, we discuss the results and conclude with the future scope of the project.

2 Previous Works

2.1 Recommendation Systems

One of the earliest techniques used to build recommendation systems is matrix factorisation [1] which splits the User-Item interaction matrix into two low-dimensional matrices that when multiplied would result in a matrix similar to the original. The incorporation of neural networks into recommender systems proves to be advantageous as the neural networks successfully capture patterns that could be missed by the simple MF technique. One such technique was the MF and NN techniques to recommend [4] to ensure that the baseline results are never worse than those produced by the MF technique.

Graph neural networks can also be used for recommendation, often modelled as a link prediction problem. Graph Convolutional Networks [5] are the most commonly used as they hold the capacity to capture both local and global relationships between nodes. A combination of graph convolution and factorization machine can be used to capture context information along with user-item interaction [6]. Social graphs can prove to be a useful component for collaborative filtering as seen in the use of [7] GNNs for the social recommendation. The usage of Graph Attention Networks has also been on the rise since these models are inherently interpretable. One can analyse the attention weights to gain useful insights about the attribute and neighbourhood importance [8]. An interesting insight into the useful components of GCNs in the recommendation scenario is that the most useful function of GCN is neighbourhood aggregation [3] which enables the foregoing of non-linear activation and feature transformation to make the model lighter and faster while maintaining the quality of recommendations. Clustering-based techniques involve clustering similar users and recommending items liked by the closest neighbours of the focal node. These techniques work even when the data is sparse and offer advantages such as diversity in recommen-dations. Methods like Artificial Bee Colony Algorithm can be used to tackle the issue caused by local optimal points in the case of the commonly used K-means clustering [2].

The content-based recommender system is the most commonly used technique to overcome the cold-start problem [3].

2.2 Evaluating Novelty in Recommendation

A Novel recommendation is defined as recommending items users have no knowl-edge about. The Novelty of a Recommender System is a metric that exists to capture user satisfaction. Accuracy on its own is not sufficient to capture the broader aspects of user satisfaction, hiding several blatant flaws in existing systems. A Recommender's ability to help user decision-making while providing new options is an essential factor in the user's satisfaction and hence the Recommender's usefulness.

2.3 Serendipity

Traditionally, Machine Learning models are been evaluated based on the classical met-rics of precision, recall, and accuracy. However, in the case of recommenders, simply recalling what the users are guaranteed to like, while increasing accuracy, may lead to boredom, saturation, and dissatisfaction. A good recommendation system gives users not only what they expect to see, but also includes some amount of unexpectedness. Novelty helps prevent the recommendations from being predictable, but it is not an adequate criterion for the discovery of un-expected, surprising, and at the same time interesting recommendations. This is captured by Serendipity. Serendipity is defined as the occurrence and development of events by chance in a happy or beneficial way [9]. Serendipity can be used as a measure of surprising a recommendation is, and avoid obvi-ous or over-specific recommendations. Due to its mostly subjective nature, Serendipity remains hard to measure experimentally.

2.4 Interpretability

Recommendation systems that are able to provide explanations for their results are often more popular among consumers as they're more transparent [10]. Inter-pretability and explainability, although frequently used interchangeably, refer to different things [11]. Interpretability refers to understanding how the model works and making the black-box nature of machine learning models more transparent, aiding the developers to better understand and fine-tune the model. Whereas, explainability refers to the technique of explaining the results obtained by the machine learning model in a way that users can understand, resulting in building user trust.

Interpretability can be achieved using two types of methods, post-hoc/model-agnostic and embedded/model-intrinsic methods. Model-intrinsic methods are limited to specific model classes. For e.g. the interpretation of regression weights in a linear model is a model-specific interpretation. Tools tailored towards neural networks are model-specific. Model-agnostic tools can be used on any machine learning model and are applied after the model has been trained (i.e., post-hoc). These methods usually work by analyzing feature input and output pairs. By definition, these methods cannot have access to model internals such as weights or structural information.

3   Data

The hybrid model is evaluated using MovieLens [12] and Last FM [13] dataset, both of which are publicly accessible.

- MovieLens: It is a commonly used dataset for movie recommendation systems. We have used the version with 20 million movie ratings from users. Although the rating values range from 1 to 5, we have scaled the ratings to binary, with a threshold rating of 3, in order to mimic implicit feedback.
- LastFM: It is a popular dataset employed for music recommendation systems. This dataset contains similarity and tag information of the million song dataset.

4   Proposed Methodology

4.1   Architecture

Figure 1 shows the outline of the hybrid recommender system. It consists of two main recommendation techniques:

- Collaborative Filtering technique
- Content-based recommendation technique

The two Collaborative filtering models are:

- Neural Collaborative Recommender System

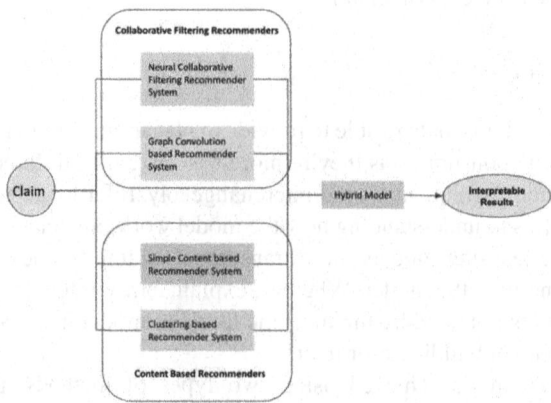

Fig. 1. Hybrid model outline.

– LightGCN / RGCN

   The two Content-based models are:

– Content-based recommender
– Cluster-based recommender

   The following subsections will explain the models in detail.

4.2 Neural Collaborative Filtering

Figure 2 shows the layers of the Neural CF RS. This model was built based on the paper [4]. It is a hybrid of two techniques: neural network-based recommendation and matrix factorization-based recommendation. The working of the model is described in the steps below.

– Two embeddings each for users and items, one for MF and one for MLP, are taken.
– The interaction matrix vectors for each user and item are concatenated and passed through a NN with non-linear activation functions.
– Simultaneously the matrix factorization technique also produces a product of the user and item embedding.
– The results for both MF and MLP are concatenated and passed through a sigmoid function.
– Finally, the loss is computed and backpropagated to fit the model.

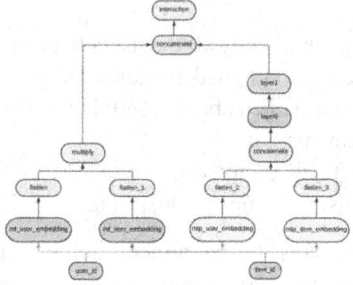

Fig. 2. Neural collaborative filtering.

4.3 Content-Based Recommender

Figure 3 shows the walkthrough of this model. The content-based recommender considers the user-assigned tags for items and performs tf-idf vectorization. Since this results in a very high-dimension vector, auto-encoders are used to reduce the dimensionality. The user vector is created by taking an average of the embeddings of the items that the user previously interacted with. The technique to construct the user and item embeddings is described below:

– Group the tags assigned to an item by all the users to construct a document.

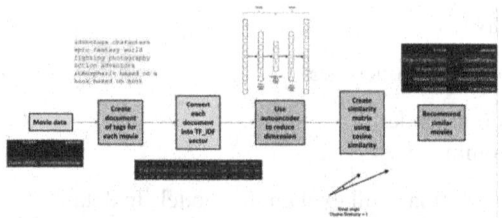

Fig. 3. Content-based walkthrough.

– Pass the document to the tf-idf vectorizer.
– Pass the tf-idf vectors to the encoder to get the final embeddings.

– To create the user embedding, an average of the embeddings of the items the user has interacted with is taken.

Finally, cosine similarity is used as a metric to calculate the distance between the user and the item vectors to recommend the items most similar to a user.

4.4 Cluster-Based Recommender

We employ a clustering-based recommendation technique where both the user and item embeddings are in the same space. The user and item embeddings are constructed as follows:

– Group the tags assigned to an item by all the users to construct that item's document. For the users, group the tags assigned to items by a user to generate that user's document. Only items that are rated above a certain threshold, or that are unrated, are considered for user documents.
– Pass the document to the tf-idf vectorizer.
– Cluster the item vectors using K-means clustering.

After getting the item clusters, to recommend items to a user, we find the closest cluster to the user and recommend 10 random items out of the ones that belong to the same cluster.

This technique enables recommending items that will match users' preferences but might be difficult for the user to find through collaborative filtering techniques that only look at the user and item neighborhood. This way we move from recommending items very similar to the user's neighborhood to the items that the user likes but faces difficulty in finding in their immediate neighborhood. Thus, this incorporates serendipity in recommendations.

4.5 Hybrid SocialLGN

This model is a derivation of SocialLGN [14], which uses LightGCN [15] as the base model and adds social user-user edges to incorporate social data into the model. Our model uses the features of the original model along with some additional techniques which have been summarised below:

- user-user implicit edges have been added to represent the social recommen-dation scenario
- item-item implicit edges have been added to represent the content-based recommen-dation scenario
- the embeddings have been initialized with the user and item embeddings constructed using tag data.

The implicit user and item edges have been constructed using a threshold of 0.5, on the similarity matrix built using the tf-idf vectors formed from the tag information representing the users and items. The embeddings have been generated by passing the tf-idf vectors to an auto-encoder that reduces the dimension to 64 which is found to be a suitable embedding size by the authors of LightGCN. This model is a hybrid of content-based (item-item edges), social (user-user edges), and collaborative filtering (user-item edges) based recommendation techniques. Thus, it benefits from the advantages of these techniques and produces promising results.

4.6 Hybrid Recommender

The hybrid recommender system combines the top recommendations from the individual models to provide the final results, with an equal weight distribution among the models. It checks the recommendations for any overlap and gives scores to the recommendations based on the number of models it was recommended by. In case of no overlap, it recommends an equal number of recommendations from each of the four models. This way a good balance in the diversity and accuracy of recommendations is maintained.

4.7 Measuring Serendipity

Serendipity in recommender systems brings in an element of pleasant surprise [16]. Our metric to calculate serendipity works in the following way:

- For each user and the recommended items to the user, we first calculate the user and item embeddings using the tag information as described in the Hybrid SocialLGN model's embedding approach.
- Then, we take an average of all the embeddings of the items recommended to the user.
- We find the cosine similarity between the user embedding and the average item embeddings calculated in the previous step.
- To calculate the serendipity, we subtract the cosine similarity score from 1.

Thus, through this distance-based metric, we can measure the element of surprise that the recommendations bring forth while maintaining their usefulness of recommendations.

4.8 Incorporating Interpretability

ELI5 To interpret the recommendations provided by the hybrid model, we have used ELI5 [7]. The users and items are passed to a multi-class classifier which classifies the users and items into a certain number of classes. This multi-class classification is interpreted by ELI5. The model provides explanations in the following way:

– If the classification of both the user and a recommended item is the same, the rec-ommendation can be explained as recommending items that are similar to your interests.
– On the other hand, if the classification of the user and the recommended item is not the same, the recommendations can be explained as a way to introduce novelty, diversity, or serendipity.

Thus, in this way, the classification problem is more interpretable and in turn, our results are more explainable. This helps the users gain more trust in the recommendation model.

WCSS Minimization To interpret the recommendations provided by the cluster-based model, we have used WCSS Minimization [17].

The K-Means clustering algorithm aims to minimize the Within-Cluster Sum of Squares (WCSS) and consequently maximize the Between-Cluster Sum of Squares (BCSS). The former measures the squared average distance of all the points within a cluster to the cluster centroid, and the latter measures the squared average distance between all centroids.

This interpretability approach is a direct analysis of each centroid's sub-optimal position. Since K-Means' aim is to minimize WCSS, we can find the dimensions that were responsible for the highest amount of WCSS minimization for each cluster by finding the maximum absolute centroid dimensional movement. Results are illustrated in Figs. 4 and 5.

Unsupervised to Supervised This approach [17] is model-agnostic, although we've used it for the cluster-based model.

In doing so, we convert the unsupervised clustering problem in the K-Means algo-rithm into a supervised classification problem using an easily interpretable classifier such as a tree-based model. The steps to do this are as follows:

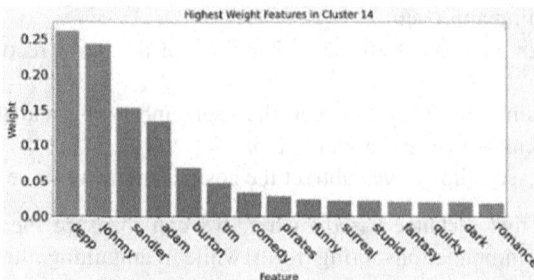

Fig. 4. Feature importance weights for Cluster 14 in the MovieLens dataset.

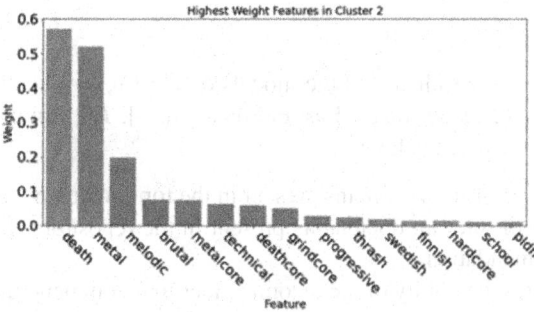

Fig. 5. Feature Importance weights for Cluster 2 in the LastFM dataset.

- Change the cluster labels into One-vs-All binary labels for each.
- Train a classifier to discriminate between each cluster and all other clusters.
- Extract the features' importance from the model. (sklearn.ensemble.RandomForestClassifier has been used in this method).

Results are illustrated in Figs. 6 and 7.

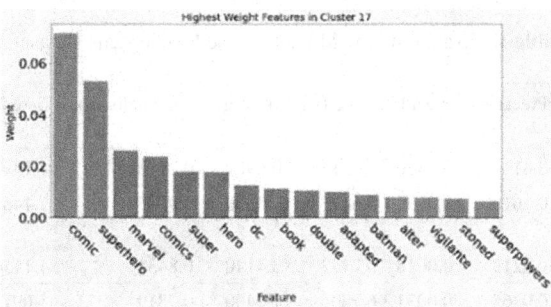

Fig. 6. Feature importance weights for Cluster 17 in the MovieLens dataset.

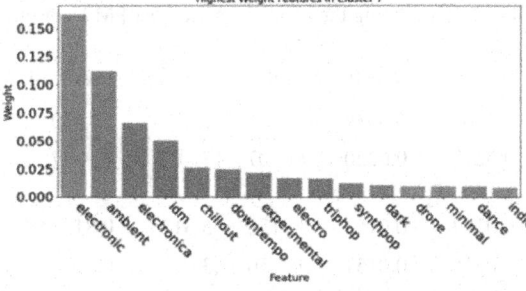

Fig. 7. Feature importance weights for Cluster 7 in the LastFM dataset.

5 Results

Table 1 summarises the results of all the models on the Movie Lens Dataset [12] and Table 2 summarises the results on the Last FM dataset [13]. The various metrics used to evaluate the model are listed below:

- Precision: Fraction of relevant items present in the top-k recommendations.
- Recall: Fraction of relevant documents present in the repository that are present in the top-k recommendation.
- Novelty: Measures the ability of the recommender to recommend items that are new to the user.
- Coverage: Fraction of items the recommender is able to recommend.
- Serendipity: Making recommendations that are both appealing and helpful requires serendipity. The advantage of this criterion over novelty and diversity is the value of useful recommendations [16].
- Intra-list similarity: Uses a feature matrix to calculate the cosine similarity between the items in a list of recommendations.
- Personalisation: Dissimilarity between the user's lists of recommendations.

The precision and recall refer to P@K and R@K where K = 10.

Table 1. Evaluating the Models on the Movie Lens Dataset.

S.No.	Model	Precision	Recall	Novelty	Coverage	Personalisation	Serendipity	Intralist Similarity
1	NN RS	0.4130	0.0329	1.4334	0.6900	0.2228	1.1398	0.2351
2	Hybrid SocialLGN	0.1610	0.0191	4.6417	16.5800	0.9837	1.1465	0.2549
3	Content	0.0215	0.0018	6.7443	3.4140	0.8434	1.1156	0.1939
4	Cluster	0.0265	0.0031	6.7804	11.9400	0.9819	1.0972	0.4709
5	Hybrid	0.1295	0.0199	4.6568	13.47	0.8113	1.2530	0.1962

Table 2. Evaluating the Models on the Last FM Dataset.

S.No.	Model	Precision	Recall	Novelty	Coverage	Personalisation	Serendipity
1	NN RS	0.2387	0.0539	10.8857	0.12	0.1430	1.1168
2	Hybrid SocialLGN	0.0213	0.0220	7.0707	17.48	0.9937	0.9822
3	Content	0.0167	0.0154	8.9132	5.3124	0.8127	0.9736
4	Cluster	0.0215	0.0157	9.7750	63.04	0.9993	1.0336
5	Hybrid	0.3705	0.5409	5.7800	34.39	0.6492	1.1055

6 Discussion

A recommendation system evaluated using metrics like precision, recall, novelty, coverage, etc. has to maintain a balance between the following contrasting goals:

- (Novelty and coverage) vs. (precision and recall) vs serendipity
- Intra-list similarity vs personalization/inter-list similarity

Table 3 summarises the value added by each of the models. The models based on collaborative filtering techniques i.e. NN RS and Hybrid SocialLGN, offer higher precision, recall, and serendipity, and the models based on content-based recommendation techniques i.e. Content-based recommender and Clustering based recommender, offer higher novelty, coverage, and personalization. Although the Hybrid SocialLGN performs well in terms of most of the metrics, it does not perform well in the cold-start scenario, thus we need to incorporate the other models in the final hybrid model. The hybrid model combines the recommendations from all four models to obtain a suitable balance of all the metrics discussed above. It offers high serendipity, novelty, and personalization score while maintaining precision and recall values.

Table 3. Comparing the Models on Contrasting Metrics

S.No.	Model	Precision & Recall	Intralist Similarity	Novelty	Coverage	Personalisation
1.	NN RS	✓				
2.	Hybrid SocialLGN	✓		✓	✓	✓
3.	Content		✓	✓		✓
4.	Cluster		✓	✓	✓	✓
5.	Hybrid	✓	✓	✓	✓	✓

7 Conclusion

A hybrid recommendation system alleviates problems like cold-start, sparse matrix, etc. in individual models and also helps in striking a balance between novelty and accuracy. Our model combines content-based, collaborative filtering-based techniques to obtain recommendations that are equal parts accurate and novel. Moreover, a new technique to build a hybrid recommender system using multiple relationship edges is explored. The overall results are made interpretable. The serendipity metric aids in assessing how unexpected the recommendations are.

8 Future Work

The interpretability of the individual models forming the hybrid model can be further investigated. The model can be evaluated on various other recommendation metrics.

References

1. Koren, Y., Bell, R., Volinsky, C.: Matrix factorization techniques for recommender systems. Computer **42**(8), 30–37 (2009)
2. Ju, C., Xu, C.: A new collaborative recommendation approach based on users clustering using artificial bee colony algorithm. Sci. World J. **2013** (2013)
3. Tai, Y., Sun, Z., Yao, Z.: Content-based recommendation using machine learning. In: 2021 IEEE 31st International Workshop on Machine Learning for Signal Processing (MLSP), pp. 1–4. IEEE (2021)
4. He, X., Liao, L., Zhang H., Nie, L., Hu, X., Chua, T.: Neural collaborative filtering. In: WWW (2017)
5. Zhang, S., Tong, H., Xu, J., Maciejewski, R.: Graph convolutional networks: a comprehensive review. Comput. Soc. Netw. **6**(1), 1–23 (2019)
6. Wu, J., et al.: Graph convolution machine for context-aware recommender system. Front. Comp. Sci. **16**(6), 1–12 (2022). https://doi.org/10.1007/s11704-021-0261-8
7. Fan, A., Jernite, Y., Perez, E., Grangier, D., Weston, J., Auli, M.: Eli5: long form question answering. arXiv preprint arXiv:1907.09190 (2019)
8. Veličković, P., Cucurull, G., Casanova, A., Romero, A., Lio, P., Bengio, Y.: Graph attention networks. arXiv preprint arXiv:1710.10903 (2017)
9. Kotkov, D., Konstan, J.A., Zhao, Q., Veijalainen, J.: Investigating serendipity in recommender systems based on real user feedback. In: Proceedings of the 33rd Annual ACM Symposium on Applied Computing, pp. 1341–1350 (2018)
10. Sinha, R., Swearingen, K.: The role of transparency in recommender systems. In: CHI'02 Extended Abstracts on Human Factors in Computing Systems (2002)
11. Li, X., et al.: Interpretable deep learning: interpretation, interpretability, trustworthiness, and beyond. Knowl. Inf. Syst. 1–38 (2022)
12. Movielens. https://grouplens.org/datasets/movielens/
13. Last.fm dataset. http://millionsongdataset.com/lastfm/
14. Liao, J., et al.: Sociallgn: light graph convolution network for social recommendation. Inf. Sci. **589**, 595–607 (2022)
15. He, X., Deng, K., Wang, X., Li, Y., Zhang, Y., Wang, M.: LightGCN: simplifying and powering graph convolution network for recommendation. In: Proceedings of the 43rd International ACM SIGIR Conference on Research and Development in Information Retrieval, pp. 639–648 (2020)
16. Ziarani, R.J., Ravanmehr, R.: Serendipity in recommender systems: a systematic literature review. J. Comput. Sci. Technol. **36**(2), 375–396 (2021)
17. K-means feature importance. https://github.com/YousefGh/kmeans-feature-importance

Network Model and Function Analysis
of Mobile Network

A Geometry-Based Strategic Placement of RISs in Millimeter Wave Device to Device Communication

Lakshmikanta Sau[(⊠)] and Sasthi C. Ghosh

Advanced Computing and Microelectronics Unit, Indian Statistical Institute,
203 B. T. Road, Kolkata 700108, India
lakshmikanta030@gmail.com, sasthi@isical.ac.in

Abstract. Recently, reconfigurable intelligent surfaces (RISs) have been introduced in millimeter wave (mmWave) device to device (D2D) communication scenarios to provide seamless connection and high data rate to a pair of proximity users. However, such high data rate can be achieved, only if the concerned device pair resides in close proximity and a direct line of sight (LoS) link exists between them. The proximity and the LoS link is necessary because of the high propagation and penetration losses of the mmWaves. The direct LoS link between a pair of devices may be blocked easily by static obstacles like buildings and trees. If there is no such direct LoS link between a pair of devices, we can use RIS to form an indirect LoS link between them. However, in that case, proper placement of RISs is necessary to provide such indirect LoS link. In this work, we develop a RIS placement strategy to serve those device pairs who do not have any direct LoS links. In order to provide an indirect LoS link for a requesting device pair, we first use some basic ideas from computational geometry to find out the candidate zones for placing RISs. Next we find the candidate zones for all such requesting device pairs considering the fact that two or more candidate zones may overlap and create a new candidate zone. We construct a graph where each candidate zone represents a vertex and there exist an edge between two overlapping candidate zones. We convert the RIS placement problem to a clique partitioning problem of the graph and use a greedy algorithm to get a near optimal solution. From simulation results, we can see that the strategically placed RISs give better performance in comparison to an existing deployment strategy, which places RISs only on the walls of the building.

Keywords: Millimeter waves · Device to Device communication · Reconfigurable Intelligent Surfaces

1 Introduction

Due to the exponential rise of mobile users, we are facing with several kind of challenges like high data rate, seamless connection, power saving and call drop in 5G device to device (D2D) communication. The millimeter wave (mmWave)

F. Neri et al. (Eds.): CCCE 2023, CCIS 1823, pp. 41–53, 2023.
https://doi.org/10.1007/978-3-031-35299-7_4

D2D communication [1] can provide high data rate in short range. But penetration loss from obstacles is a major problem in mmWave D2D communication. Penetration losses [2] happen due to the randomly located obstacles like trees and buildings in an urban setup. Because of the obstacles present in the environment, providing a direct line of sight (LoS) link for a pair of devices is a challenging task. An indirect LoS link may be established by avoiding obstacles by using reconfigurable intelligent surfaces (RISs). RISs are very useful in mmWave D2D communication technology [3], which can change a propagation environment into a desired form [4]. RISs consist of a large number of reflecting elements, which can shift the phase of an incident signal into a desired direction. By this way, RIS can provide an indirect LoS path for blocked links. In general, RISs are cost and energy efficient. So by deploying such software controlled programmable RISs, we can extend the range of wireless communication significantly. However, to achieve such a short indirect LoS link for a communicating device pair, a proper placement of RISs is necessary. Our main aim is to develop a strategic placement RISs so that an indirect LoS link may be provided for the device pair whose direct LoS is blocked by obstacles.

In a recent study in [5], a large number of RISs are placed on the walls of the randomly located obstacles like buildings. Here all the RISs are strategically placed on the wall of the obstacles so that coverage area can be improved significantly. In [5] obstacles are considered as a line segment and RISs are coated in one of the side of the line segments. Such placement of RISs can leave many pairs of devices remain uncovered, as demonstrated in Fig. 1. Note that in Fig. 1 there is no direct LoS link between w and v because of the presence of the building B. But an indirect LoS link can be provided between w and v using the RIS deployed on the wall of the building C. It is evident that no such indirect LoS link can be established between u and v even using the RIS coated on the walls of the two buildings C and B. An RIS placed in a strategic place (marked as A) can easily provide an indirect LoS link between u and v.

The main focus of this work is to place RISs in the strategic locations so as to increase the coverage area significantly. More specifically, we aim to develop a strategy for the deployment of RISs in the service area using a computational geometric approach. The challenge of avoiding obstacles in the environment is a major issue which once solved can greatly improve the performance of the D2D networks. Our main concern for the strategic deployment of RISs is for those device pairs which do not have direct LoS link. To solve this RISs placement problem, we consider a pair of devices which do not have direct LoS between them (termed as blind pair) and then find out a common region if any which is visible from both the devices. In that case an RIS can be placed there to serve the concerned blind pair. Next, we consider all the blind pairs that exist in the service region. Note that there can be overlapping regions which are visible to multiple blind pairs. After finding the overlapping regions, we will construct a graph G where each candidate zone represents a vertex and two vertices will have an edge if the concerned candidate zones overlap.

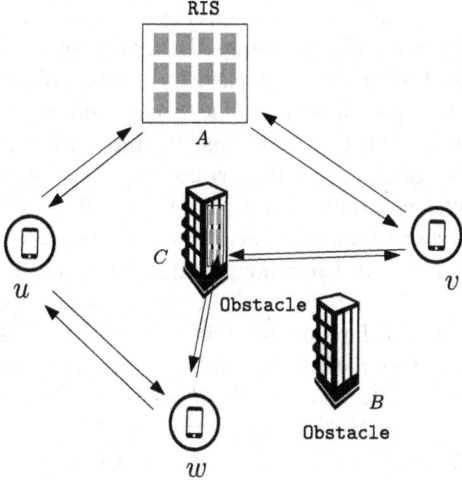

Fig. 1. RIS assisted communication

The RIS placement problem is then converted into the clique partitioning problem of the graph G. Finally, RISs will be installed based on the near optimal solution of the clique partitioning problem. Our objective will be to place a minimum number of RIS to serve all such blind pairs. To the best of our knowledge, this is the first work which provide a framework to improve the coverage areas by deploying RISs in strategic locations in a continuous domain.

This paper is organized as follows. In Sect. 2, we have discussed about the related and existing works. In Sect. 3, we have clearly discussed the system model and communication scenario considered. In Sect. 4, we present the strategical deployment process of RISs in step by steps. In Sect. 5 we have discuss about the simulation results and compare the simulated results with an existing RISs deployment strategy. Finally, in Sect. 6 we give concluding remarks and future direction of the research.

2   Related Works

There are several works in mmWave D2D wireless communication studying the impact and benefits of using RISs. The authors in [6] gave a clear concept on software defined RISs. In [5] authors gave an idea about the strategic placement of RISs on the wall of the randomly located obstacles like buildings. When a device want to connect to another device then they have to select a better channel by which they can communicate to each other. In [7–10], authors have discussed about the channel selection strategies in mmWave D2D communication. Path loss is a main issue in mmWave D2D communication. In [11,12] authors gave a clear clarification on path loss and how other parameters affect it. The authors of [13] provided a clear strategy for increasing bandwidth performance. In [5], the authors improve the coverage areas in cellular networks by installing the RISs on the wall of the obstacles like buildings. In mmWave indoor communication scenario [14], the LoS link may be blocked by various obstacles like buildings and furniture. To overcome these obstacles in an indoor scenario, RISs have been installed on the wall of the buildings. In mmWave D2D communication, energy efficiency is one of the most important issue. The idea of exploiting and installing RISs in D2D communication for energy efficiency has been studied in [15]. Here also authors assume that RISs are placed randomly inside the service area. From [16] we can learn how localization accuracy can be improved by deploying large number of RISs. In [17], the authors have discussed how to reduce the aggravated interference brought on by D2D links using the same resources. In [18] the authors study how to maximize the sum rate for a pair of devices under specific conditions, such as an outage probability and a signal-to-interference-plus-noise ratio (SINR) target threshold.

From the above discussion we can see that the uses of RISs bring a renaissance in the field of mmWave D2D communication. More specifically, RISs are used to bypass signal blockage, aid in eliminating interference and raising the system sum rate for D2D communication. However, most of the studies assume that RISs are deployed either on a random location or in a strategic location on the wall of the building. For random deployment of RISs, it may be cost inefficient and that may lead to many RISs which are not necessary. Placing RISs only the walls may leave many pairs of devices remain uncovered, as demonstrated in Fig. 1. So, we have to deploy RISs in a strategical way such that we can cover maximum area using less RISs. Strategical placement of RISs also help to increase the throughput and user coverage in D2D communication. Instead of placing RISs on the walls only, in this work, we consider deploying RISs in any strategic location within the service area using a computational geometric approach.

3   System Model

In this work, we consider the system model similar to [5]. We consider a D2D communication scenario (Fig. 1), where the devices/users are modelled as independent homogeneous Poisson point processes (PPPs) $\Psi_u = \{u_i\}$ $i \in R^2$ with density λ_u. Here we consider the blockage as Boolean model. We consider that a blockage is a line segment with length L_i and angle θ_b. The midpoint location of the line segments representing blockages are model as a PPPs $\Psi_b = \{z_i\}$ $i \in R^2$ with density λ_b. The line segment L_i is uniformly distributed over a bounded interval (L_{min}, L_{max}). Angle between the positive x-axis and the line segment L_i is $\theta_{b,i}$ where it is uniformly distributed over 0 to 2π. We assumed that there are n device pairs which do not have direct LoS due to the presence of obstacles. Mainly we will focus on to place RISs in strategic locations such that these n device pairs which do not have direct LoS links can be served by providing indirect LoS links via the RISs. Here we want to minimize the number of RISs as an unnecessary RIS may cause interference to other device pairs. Now we will state two definitions which will be used throughout the manuscript.

Definition 1 (Direct LOS link [5]). *If two users u and v reside within r distance and there is no blockage between them then the link between them is called a direct LOS link.*

Definition 2 (Indirect LOS link [5]). *Two users u and v residing within 2r distance is said to have an indirect LoS link between them, if u and v do not have a direct LOS link between them but there exist an RIS s such that both the link $u \to s$ and $s \to v$ have direct LOS.*

In the following subsections, we first introduce the path loss model and channel model for RIS-assisted D2D communication.

3.1 Path Loss Model

In order to evaluate the efficiency of RISs deployment, we have to state an appropriate path loss model for a received signal through an indirect link via a RIS. In other words, we have to measure the signal gain in the receiver end for such an indirect communication. In [5], authors have provided a clear concept about the relationship between the received signal power and the distance traveled through the indirect path. Let M is the number of meta surfaces of each RIS. Let i^{th} transmitter transmit a signal and j^{th} receiver will receive this signal via a RIS. Then the power of received signal at j^{th} receiver is scaled by $M^2(d_{i-R} + d_{R-j})^{-\alpha}$ where d_{i-R} is distance between i^{th} user and RIS R, d_{R-j} is the distance between RIS R to j^{th} user and α is path loss exponent. If there is a direct LoS link between i^{th} user and j^{th}, then power received at j^{th} user is $d_{i-j}^{-\alpha}$ where d_{i-j} is the distance between i^{th} user and j^{th} user.

3.2 Channel Model and System Throughput

We consider the channel model similar to [7]. Suppose there are n device pairs who want to communicate with each other at a particular time. We assume that at a particular time instant, a device can communicate with exactly one device. Let in a particular time, u_1 want to connect with v_1. Then we consider that (u_1, v_1) is a requesting pair. Let us consider that collection of all such requesting pair is Z. So, $Z = \{(u_1, v_1), (u_2, v_2) \ldots (u_n, v_n)\}$. The channel coefficient between u_i and v_i is $h_{i,i} \in \mathbb{C}$. If there is a direct LOS link between u_i and v_i, then they can connect each other directly. If there is no such direct LOS link, then they will be connected by an indirect LOS link by using an RIS. Assume S as the set of RISs and each RIS has M meta surfaces. Then there will have a channel between u_i and s^{th} RIS. This channel vector is denoted as $\mathbf{f}_i^s \in \mathbb{C}^{M \times 1}$. Also there exist a channel vector from s^{th} RIS to v_i which is denoted as $\mathbf{g}_j^s \in \mathbb{C}^{M \times 1}$. Total number of reflecting element is given by $P = M \times |S|$. Hence channel coefficient vectors from the u_i to all the RISs and from all the RISs to v_i are $\mathbf{f}_i = (\mathbf{f}_i^1, \mathbf{f}_i^2, \mathbf{f}_i^3, \mathbf{f}_i^4, \ldots \mathbf{f}_i^S) \in \mathbb{C}^{P \times 1}$ and $\mathbf{g}_i = (\mathbf{g}_i^1, \mathbf{g}_i^2, \mathbf{g}_i^3, \mathbf{g}_i^4, \ldots \mathbf{g}_i^P) \in \mathbb{C}^{P \times 1}$ respectively [15]. Let $\Theta = \sqrt{\eta} \, diag(\theta)$ is a phase shift matrix of the RISs, where $\theta = [\theta_1, \theta_2, \theta_3, \ldots \theta_P]^H$ and η lie between $(0, 1)$. Here Θ is a diagonal matrix.

For a communication link between user u_i and v_i, let u_i transmit a signal and v_i will receive the transmitted signal. Let transmit power of u_i is P_{u_i}. Then the SINR of the link between u_i and v_i can be expressed as [7,15]

$$z_i = \frac{|(h_{ii} + \mathbf{g}_i^H \Theta \mathbf{f_i})|^2 P_{u_i}}{\sum_{\substack{l=1 \\ l \neq i}}^{n} |h_{li} + \mathbf{g}_i^H \Theta \mathbf{f}_l|^2 P_{u_i} + \sigma^2} \tag{1}$$

where σ^2 is the Gaussian noise factor. Now from the Shannon's capacity formula we shall get the throughput for a device pair (u_i, v_i) and adding the throughput of all such pairs, we will get the system throughput.

4 Strategic Deployment of RIS

In this section, we will discuss about the strategical deployment of RISs. Here we first consider a single device pair which do not have a direct LoS between them and find out the candidate zones where we can deploy the RISs such that an indirect LoS link can be established between them. Next we will find out the candidate zones for all such device pairs which do not have direct LoS between them. Note that the candidate zones of different device pairs may overlap. Next we will find out such overlapping zones considering all the device pairs and finally we will use a clique partitioning approach to install RISs. In the following subsections, we will discuss about these steps in details.

4.1 Candidate Zone for a Single Device Pair

Let a device can be connected directly with other device if they are within a specific distance r and have a direct LoS between them. Consider two devices u and v residing within $2r$ distance. Let u and v do not have a direct LoS link between them. In this case an RIS may help establishing an indirect LoS between them.

Let O_1 be the position of u and O_2 be the position of v. Let us consider a circle C_1 with centre at O_1 and radius r. Also, let us consider a circle C_2 of radius r with centre at O_2. An RIS may be deployed in the common intersecting zone of C_1 and C_2 in order to provide an indirect LoS between u and v. Let A and G be the intersecting points of C_1 and C_2. Let C_{12} denote the intersection zone of C_1 and C_2. Join A with both O_1 and O_2. Similarly join G with both O_1 and O_2. After joining these points we will have a quadrilateral O_1AO_2G. Now we will consider all those obstacles whose at least one end point is within the quadrilateral O_1AO_2G. Let L_1 be the set of endpoints of the line segments whose at least one end point is within the sector AO_1G of C_1. Also, let L_2 be the set of endpoints of the line segments whose at least one end point is within the sector AO_2G of C_2. Let us define S_1 as the set of all straight lines which passes through O_1 and an end point in L_1. Similarly define S_2 as the set of all straight lines

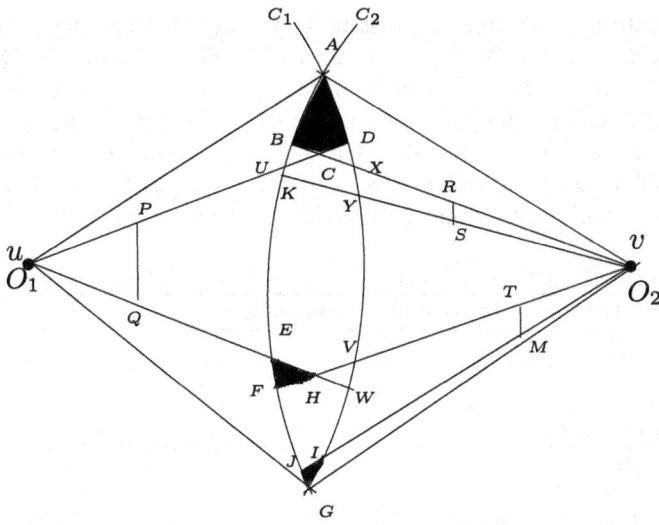

Fig. 2. candidate zone

which passes through O_2 and an end point in L_2. Let $L = S_1 \cup S_2$. Each straight line in L meets the arc AG of C_1 as well as the arc AG of C_2. Let $U_1 = \{x_i :$ the straight line $O_1 l_i$ meets the arc AG of C_1 at x_i and $l_i \in L_1\}$ and $U_2 = \{y_i :$ the straight line $O_1 l_i$ meets the arc AG of C_2 at y_i and $l_i \in L_1\}$. Similarly let us define $V_1 = \{x_i :$ the straight line $O_2 l_i$ meets the arc AG of C_1 at x_i and $l_i \in L_2\}$ and $V_2 = \{y_i :$ the straight line $O_2 l_i$ meets the arc AG of C_2 at y_i and $l_i \in L_2\}$. Define $I = \{z : s_1 \in S_1$ and $s_2 \in S_2$ meet at z and z is inside $C_{12}\}$. Now, let us consider that $S = U_1 \cup U_2 \cup V_1 \cup V_2 \cup I$. A point $x \in S$ is said to be *visible* if it is visible from both the points O_1 and O_2. In other words, both the straight lines xO_1 and xO_2 do not intersect any line segment in S_1 and S_2 respectively. A point $x \in S$ is said to be *invisible* if it is not visible. Let $S^v = \{x \in S : x$ is visible$\}$ be set of all visible points in S. We now sort all the points of S^v with respect to their y coordinates. A zone is said to be a *candidate zone* if all points in that zone are visible from both O_1 and O_2. Let PQ be an obstacle. Let $O_1 = (0,0)$ be the origin. Let $P = (x_1, y_1)$ and $Q = (x_2, y_2)$ be the two end points of the line segment PQ. The line $O_1 P$ is said to be an *upper arm* or a *lower arm* if $y_1 > y_2$ or $y_1 < y_2$ respectively. If $y_1 = y_2 < 0$ then $O_1 P$ is said to be an upper arm or a lower arm if $x_1 > x_2$ or $x_1 < x_2$ respectively. If $y_1 = y_2 > 0$ then $O_1 P$ is said to be an upper arm or a lower arm if $x_1 < x_2$ or $x_1 > x_2$ respectively. If $y_1 = y_2$ and $x_1 = x_2$ then PQ is a point obstacle.

Our aim is to partition the set of visible points in S^v into disjoint subsets such that the points in each subset will form a candidate zone. We now explain the process to construct the candidate zones. Consider any line segment $l \in L$. The line segment l will partition the set of points in S^v into two disjoint subsets G_1 and G_2 such that all the points in G_1 is above the upper arm of l and all the points in G_2 are below the lower arm of l. Similarly, we can obtain two disjoint subsets for each line segment in L. Finally this will partition S^v into some k disjoint subsets $P_1, P_2, \cdots, P_k$ such that all the points in P_i is either above or below of any line segment in L. In Algorithm 1, we will use S^v and L as input and it will produce the subsets $P_1, P_2, \cdots, P_k$. Now applying Graham scan algorithm on each P_i we will get a convex polygon which will eventually form a candidate zone. Note that Algorithm 1 uses Algorithm 2 as a function to determine whether a given point is below a given line or not.

Algorithm 1: Candidate zone finding algorithm

   Input: S^v, L

   Output: R

1 Mark first point of S^v as begin;
2 mark last point of S^v as end;
3 **for** l *in* L **do**
4 | **for** x *in* S^v **do**
5 | | **if** *Is_below(x,l)* **then**
6 | | | mark x as begin;
7 | | | mark prev(x) as end;
8 | | | break;
9 | | **end**
10 | **end**
11 **end**
12 m=0;
13 **for** x *in* S^v **do**
14 | **if** *x is not marked end* **then**
15 | | store x in P_m
16 | **else**
17 | | store x in P_m;
18 | | created P_m
19 | **end**
20 | m=m+1;
21 **end**
22 Return family $R = \{P_1, P_2, \ldots, P_{m-1}\}$

We now demonstrate Algorithm 1 using Fig. 2. Let u and v be two devices and they want to communicate with each other. Let PQ, RS and TM be three obstacles lie between u and v as shown in Fig. 2 and also they are within the quadrilateral O_1AO_2G. Here RS and TM lie within sector AO_2G and PQ lies within sector AO_1G. Here $L_1 = \{P, Q\}$ and $L_2 = \{R, S, T, M\}$. Here A and G are the intersecting points of C_1 and C_2. So, from Fig. 2 we have $U_1 = \{D, W\}$, $U_2 = \{U, E\}$, $V_1 = \{X, Y, V, I\}$, $V_2 = \{B, K, F, J\}$, $I = \{C, H\}$, $S = \{A, B, D, C, X, U, K, Y, E, V, H, W, F, I, J, G\}$ and $S^v = \{A, B, D, C, E, H, F, I, J, G\}$. Here we get $L = \{O_1P, O_1Q, O_2R, O_2S, O_2T, O_2M\}$. Here S^v is partitioned into three subsets $P_1 = \{A, B, C, D\}$, $P_2 = \{E, F, H\}$ and $P_3 = \{I, J, G\}$. The convex polygon representing the candidate zones based on P_1, P_2 and P_3

Algorithm 2: *Is_below*

Input: S^v, L
Output: True of False
1 $Is\_below(x \in S^v, l \in L)$;
2 **if** *l is lower arm* **then**
3 | **if** *x is on or below l* **then**
4 | | return True
5 | **else**
6 | | return False
7 | **end**
8 **if** *l is upper arm* **then**
9 | **if** *x is on or below l* **then**
10 | | return False
11 | **else**
12 | | return True
13 | **end**

---

as obtained by applying Graham Scan algorithm are shown in Fig. 2 as black colored zones. So, if we install an RIS in any of the candidate zones formed by either of P_1, P_2, or P_3 then we can serve u and v by an indirect LoS link via an RIS.

The worst case time complexity is $O(k^3)$ for finding the candidate zones for single device pair, where k is the total number of obstacles.

4.2   Candidate Zones for All Device Pairs

In the previous section we have discussed the scenario for one device pair u_1 and v_1 in the D2D mobile communication scenario. Now we will find out the candidate zones for all the device pairs who are not connected by a direct LoS link. Let there be n such device pairs. For every such device pair (u_i, v_i), there may be one or more candidate zones which are convex polygons and are stored in the set P_{u_i,v_i}. Let $\mathbf{R} = \{P_{u_1,v_1}, P_{u_2,v_2}, ...P_{u_n,v_n}\}$ be the collection of all the candidate zones. There may even exist few device pairs whose candidate zones overlap with each other. In this work, our main challenge is to find out the overlapping regions between candidate zones of $\mathbf{R}$.

We denote $\mathbf{L_R}$ to be the set of all line segments forming the boundaries of all the convex polygons of $\mathbf{R}$. To find out the overlapping zones, we will sort L_R with respect to the x coordinate and referenced to the polygon they belong to. After sorting $\mathbf{L_R}$ we will apply line sweep [19] algorithm on $\mathbf{L_R}$. Seeing the status of the vertical sweep line we can determine which polygon are neighbour to each other. We check for the intersection of the neighbouring polygons and we store the intersection regions if any. To find out the overlapping zones we will use Algorithm 3. We use $\mathbf{L_R}$ and $\mathbf{R}$ as inputs of Algorithm 3 and it will produce the overlapping candidate zones $\mathbf{Z_R}$ as outputs.

The worst case time complexity is $O(nk^2 \log(nk^2))$ for finding the candidate zones for all the n device pairs, where k is the total number of obstacles.

Algorithm 3: Polygon intersection finding

Input: R, L_R
Output: Z_R

1 Sort L_R with respect to x coordinate;
2 Reference the line segment in L_R with respect to the polygon it belongs to;
3 Run line sweep from left to right;
4 Using line sweep check intersection of the adjacent polygon and store the corresponding intersection zone in Z_R.
5 Return Z_R

4.3   Clique Based RIS Placement Strategy

Let m be the total number of candidate zones in Z_R. We construct a graph G with m vertices where each candidate zone is represented by a vertex and two vertices will have an edge if their representing zones overlap. Here we want to divide the vertices of G into minimum number of disjoint clusters, where each cluster forms a clique. Then, we check the intersecting regions for each clique and install an RISs anywhere in this region. Such an RIS will serve all the device pairs corresponding to the candidate zones forming the clique. For this step we use the near optimal solution of the clique partition problem [20].

It may be noted that even after doing this work, there may have few device pairs who are not cover by any RIS.

5   Simulation Results

We consider simulation parameters similar to [7]. We consider a $1000m \times 1000m$ square area. We assume that the threshold distance for direct LoS is $60m$. In other words, we assume that if the receiver is more than $60m$ from transmitters or RISs then it does not receive signal with enough power for the required SINR threshold. Also we consider 60 GHz frequency and 500 MHz bandwidth for D2D communication [21]. The RIS reflect efficiency is assumed to be $\eta = 0.8$. As in [17], we consider that a device transmission power is 23 dBm. We use Rician fading channel with β Rician factor. Here β tends to infinity. We assumed that for obstacle free path, path loss exponent $\alpha = 1.88$ and SINR threshold is 15 dB as used in [22]. The additive Gaussian noise $\sigma^2 = -117$ dBm/Hz is taken as in [15].

We now compare our proposed strategy with the strategy where RISs are placed only on the walls [5, 23–25]. Here we compare the performance of the above two strategies with respect to system throughput, outage percentage and number of required RISs. Here system throughput is defined as the sum throughput obtained from the individual links as detailed in Sect. 3.2. Outage percentage represents the percentage of requesting links which are not served.

In Fig. 3, number of RISs increases with respect to the rising of obstacles. This is because due to rising of obstacle, most direct LoS link are blocked. In

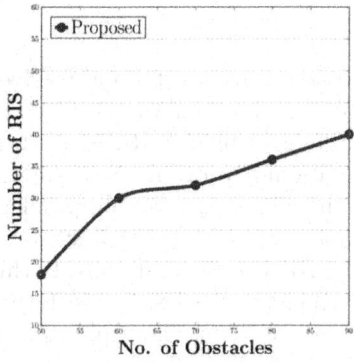

Fig. 3. No. of obstacles vs. RIS

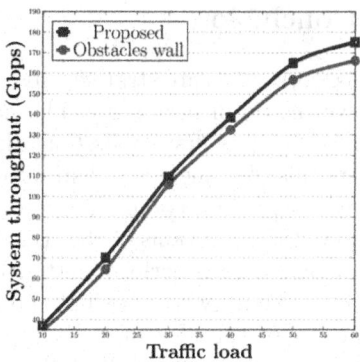

Fig. 4. Traffic vs. System throughput

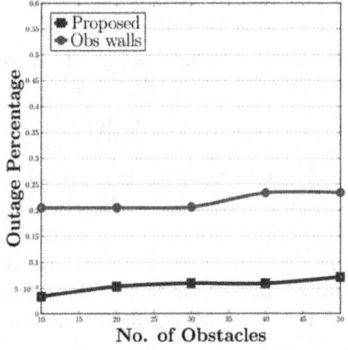

Fig. 5. No. of obstacles vs. Outage Percentage

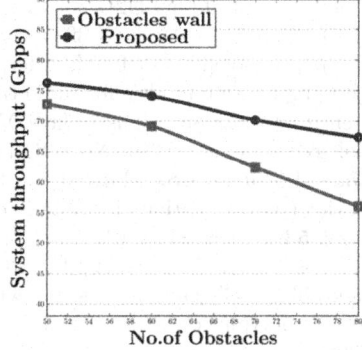

Fig. 6. No.of obstacles vs. System throughput

Fig. 4, the system throughput increases with traffic load up to certain point and then saturates in both cases. However better throughput is obtained by the proposed strategy. Here number of requesting device pairs vary from 10 to 60 and they are selected randomly. In Fig. 5, outage percentage increases with number of obstacles, as expected. However, our proposed strategy produces less outage percentage than the strategy where RISs are deployed on the walls. Here number of obstacles vary from 50 to 80. In Fig. 6, the system throughput decrease as the number of obstacle increase. Our proposed deployment strategy gives better system throughput with respect to the deployment of RISs on the walls. Here number of obstacles vary from 50 to 80 and they are deployed randomly inside the region. So, overall our proposed strategy produces better results with respect to system throughput and outage percentage than that of deploying RISs on the wall of the obstacles.

6   Conclusion

In this work, RISs are strategically deployed between a transmitter and receiver to solve the problem of signal blockage due to obstacles in mmWave D2D communication. Here first we find out the candidate zone for placing RISs and then convert this placement problem into a clique partitioning problem. Next we use a greedy algorithm commonly used for solving the clique partitioning problem to find out the positions of the RISs. The experiment result shows that strategically deployed RISs perform better than deploying RISs on the walls only. In this work, we mainly work on single hope RISs placement strategies, but in future we can deploy the RISs in multi hop scenario, where we can use multiple RISs in between two devices such that the coverage area can be improved significantly.

References

1. Ansari, R., et al.: 5G D2D networks: techniques, challenges, and future prospects. IEEE Syst. J. **12**(4), 3970–3984 (2017)
2. Hoppe, R., Wolfle, G., Landstorfer, F.: Measurement of building penetration loss and propagation models for radio transmission into buildings. In: Gateway to 21st Century Communications Village. VTC 1999-Fall. IEEE VTS 50th Vehicular Technology Conference (Cat. No. 99CH36324), vol. 4, pp. 2298–2302. IEEE (1999)
3. Qiao, J., Shen, X.S., Mark, J.W., Shen, Q., He, Y., Lei, L.: Enabling device-to-device communications in millimeter-wave 5G cellular networks. IEEE Commun. Mag. **53**(1), 209–215 (2015)
4. ElMossallamy, M.A., Zhang, H., Song, L., Seddik, K.G., Han, Z., Li, G.Y.: Reconfigurable intelligent surfaces for wireless communications: principles, challenges, and opportunities. IEEE Trans. Cognit. Commun. Networking **6**(3), 990–1002 (2020)
5. Kishk, M.A., Alouini, M.-S.: Exploiting randomly located blockages for large-scale deployment of intelligent surfaces. IEEE J. Sel. Areas Commun. **39**(4), 1043–1056 (2020)
6. Liaskos, C., Tsioliaridou, A., et al.: Initial UML definition of the hypersurface programming interface and virtual functions. Europ. Commiss. Project VISORSURF: Accept. Public Deliverable D **2**, 18008–18011 (2017)
7. Deb, S., Ghosh, S.C.: An RIS deployment strategy to overcome static obstacles in millimeter wave D2D communication. In: 2021 IEEE 20th International Symposium on Network Computing and Applications (NCA), pp. 1–8. IEEE (2021)
8. Chen, J., Liang, Y.-C., Cheng, H.V., Yu, W.: Channel estimation for reconfigurable intelligent surface aided multi-user MIMO systems. arXiv preprint arXiv:1912.03619 (2019)
9. Raj, S.M.G., Bala, G.J., Sajin, M.M.: 2D discrete cosine transform based channel estimation for single user millimeter wave communication system. J. Commun. **15**(2), 205–213 (2020)
10. Dong, J., Zhang, W., Yang, B., Sang, X.: WSDSBL method for wideband channel estimation in millimeter-wave MIMO systems with lens antenna array. J. Commun. **15**(11), 826–832 (2020)
11. Di Renzo, M., et al.: Reconfigurable intelligent surfaces vs. relaying: differences, similarities, and performance comparison. IEEE Open J. Commun. Soc. **1**, 798–807 (2020)

12. Tang, W., et al.: Path loss modeling and measurements for reconfigurable intelligent surfaces in the millimeter-wave frequency band. IEEE Trans. Commun. **70**, 1 (2022)
13. Sandi, E., Rusmono, A.D., Diamah, A., Vinda, K.: Ultra-wideband microstrip array antenna for 5G millimeter-wave applications. J. Commun. **15**(2), 198–204 (2020)
14. X. Tan, Sun, Z., Koutsonikolas, D., Jornet, J.M.: Enabling indoor mobile millimeter-wave networks based on smart reflect-arrays. In: IEEE INFOCOM 2018-IEEE Conference on Computer Communications, pp. 270–278. IEEE (2018)
15. Jia, S., Yuan, X., Liang, Y.-C.: Reconfigurable intelligent surfaces for energy efficiency in D2D communication network. IEEE Wireless Commun. Lett. **10**(3), 683–687 (2020)
16. He, J., Wymeersch, H., Kong, L., Silvén, O., Juntti, M.: Large intelligent surface for positioning in millimeter wave MIMO systems. In: 2020 IEEE 91st Vehicular Technology Conference (VTC2020-Spring), pp. 1–5. IEEE (2020)
17. Chen, Y., et al.: Reconfigurable intelligent surface assisted device-to-device communications. IEEE Trans. Wireless Commun. **20**(5), 2792–2804 (2020)
18. Cai, C., Yang, H., Yuan, X., Zhang, Y.-J.A., Liu, Y.: Reconfigurable intelligent surface assisted D2D underlay communications: a two-timescale optimization design. J. Commun. Inf. Networks **5**(4), 369–380 (2020)
19. De Berg, M.C.: "Kreveld m. and overmars m. 2008." Computational Geometry Algorithms and Applications 3rd Ed. Springer-Verlag. de Berg M. Cheong O. Kreveld M. and Overmars M (2008). https://doi.org/10.1007/978-3-540-77974-2
20. Dessmark, A., Jansson, J., Lingas, A., Lundell, E.-M., Persson, M.: On the approximability of maximum and minimum edge clique partition problems. Int. J. Found. Comput. Sci. **18**(02), 217–226 (2007)
21. Al-Hourani, A., Chandrasekharan, S., Kandeepan, S.: Path loss study for millimeter wave device-to-device communications in urban environment. In: 2014 IEEE International Conference on Communications Workshops (ICC), pp. 102–107. IEEE (2014)
22. Singh, D., Ghosh, S.C.: Mobility-aware relay selection in 5G D2D communication using stochastic model. IEEE Trans. Veh. Technol. **68**(3), 2837–2849 (2019)
23. Peng, Z., Li, T., Pan, C., Ren, H., Xu, W., Di Renzo, M.: Analysis and optimization for RIS-aided multi-pair communications relying on statistical CSI. IEEE Trans. Veh. Technol. **70**(4), 3897–3901 (2021)
24. Nemati, M., Park, J., Choi, J.: RIS-assisted coverage enhancement in millimeter-wave cellular networks. IEEE Access **8**, 188171–188185 (2020)
25. Chen, Y., Wang, Y., Zhang, J., Li, Z.: Resource allocation for intelligent reflecting surface aided vehicular communications. IEEE Trans. Veh. Technol. **69**(10), 12321–12326 (2020)

Obstacle Aware Link Selection for Stable Multicast D2D Communications

Rathindra Nath Dutta[✉][iD] and Sasthi C. Ghosh[iD]

Advanced Computing and Microelectronics Unit, Indian Statistical Institute,
Kolkata, India
ratcoinc@gmail.com, sasthi@isical.ac.in

Abstract. The rapid growth of multimedia applications requiring high
bandwidth has paved the way for millimeter-wave (mmWave) device-to-
device (D2D) communications. In many modern applications, such as
video streaming, same data packets need to be delivered to a group of
users. Multicasting these packets has a clear advantage over repeated
unicast. Establishing a stable multicast route for mmWave D2D communications is a challenging task as presence of obstacles can easily break
a link. In this work we devise a mechanism for constructing a stable
route for D2D multicast communications in presence of static as well
as dynamic obstacles. This requires some knowledge about the blockage
probabilities due to the obstacles. We then present a way of learning
these blockage probabilities using the Dempster-Shafer evidential theory
framework. We show the effectiveness of our proposed scheme over an
existing approach through extensive simulations.

Keywords: Device-to-device communications · Millimeter-wave
signal · Obstacles · Multicasting · Evidential theory

1 Introduction

The explosion of mobile devices together with the applications demanding high
bandwidth have already saturated the conventional wireless communication systems. In next generation wireless communication, new strategies such as millimeter wave (mmWave) device-to-device (D2D) communication has been proposed to satisfy the high bandwidth requirements [1]. The D2D communication
is the enabling technology that allows two user equipments (UEs) within proximity of each other, to directly communicate with each other with limited or
no involvement of the base station (BS). Such short distance communication
using mmWave not only provides higher received signal strength but also limits
the interference to other UEs communicating using the same frequency channel [1]. Furthermore, using beamforming technique one can send highly directed
mmWave signals and achieve pseudo-wire like communication [2]. Since D2D
communication using mmWave signals incurs very limited interference, it allows
us to activate more than one D2D communication links in the same frequency

F. Neri et al. (Eds.): CCCE 2023, CCIS 1823, pp. 54–66, 2023.
https://doi.org/10.1007/978-3-031-35299-7_5

channel, which enhances the spectral efficiency and in turn raises the overall system throughput [3,4]. One drawback of using such high frequency mmWave signals is that they are more susceptible to propagation and penetration losses due to their smaller wavelengths [5]. Therefore, it requires short distance obstacle free line-of-sight (LOS) communication path between the transmitter-receiver pair in order to achieve the promised high data rate [6]. Note that, here an obstacle can be anything from brick wall, signboard to moving objects like an automobile or can even be a person [7,8].

Many modern applications such as video streaming, automotive, IoT, public safety systems requires same data packets to be delivered to multiple client devices [9,10]. In many cases such messages are to be sent only to a selected small group of users in the network. In such cases, flooding the entire network in order to broadcast such data packets is not desirable and gave rise to the study of multicast techniques [11], where a transmitter can simultaneously send data to more than one receiver [10]. The ability of multicasting has already been considered in LTE-A specs and also being considered for D2D communication in 5G [12]. Given a source and set of destination devices, a multicast route is first determined, following which the packets are transmitted. Typically, such multicast routes forms a spanning tree rooted at the source device [12]. Multicasting using mmWave D2D communications has its own challenges. Different links in a multicast route are of different lengths and may suffer from different channel conditions due to severe propagation loss of mmWave signals. Apart from this heterogeneous links, the presence of obstacles further varies the link quality [3,9,13] due to severe penetration losses of mmWave signals from obstacles. Thus, all the links forming a multicast route, must be judiciously chosen.

In order to establish a communication link between a transmitter T_x and a receiver R_x, it must be assigned a frequency channel and a transmit-power must also be specified at T_x. Since the number of frequency channels are limited and usually smaller than the number of users, the channel must be cooperatively shared among the users. The link selection and associated resource allocation problem for multicast D2D communications has been well studied. The authors of [14] deals with power control for multicasting in vehicle-to-anything (V2X) communications. Both the authors of [15,16] presents a power control scheme for multimedia broadcast multicast services (MBMS) of LTE standards. In [10], the authors studies the channel allocation problem for multicast D2D communications. A relay aided D2D multicasting has been explored in [9] for content delivery in social networks.

In a multicast network, the overall performance of the system relies on all the selected communication links forming the multicast route. Thus, presence of obstacles, specially dynamic obstacles, has a much greater impact on the overall system performance. Although, the obstacle aware link selection and resource allocation problem has been well investigated for D2D unicast communications [4,8]. While most of the existing studies for multicast D2D communications does not consider the presence of any obstacles, a very few recent studies propose obstacle aware frameworks for D2D multicast communications. In [7], the authors have proposed a D2D-aided multicast solution in order to avoid any blockage due to presence of any obstacle considering a fixed mobility model for the dynamic obstacles. In [17], the authors have used a radar mounted at the BS to track the movement of large dynamic obstacles and analyzes the blockage detection probability with stochastic geometry. While in [18], the authors proposed deployment of cameras in the service area for detecting human movements which is then used to select obstacle free D2D links.

Most of the existing obstacle aware frameworks either assumes a fixed known distribution of the dynamic obstacles [7] or utilizes some extra hardware such as radar or camera to detect obstacle motion [17,18]. These solutions cannot dynamically adapt the change in the environment or are too expensive. Instead, in this work, we present an inexpensive solution to dynamically learn the blockage patterns due to the dynamic obstacles. We then apply it for a stable route selection for multicast D2D communications so that the link breakages due to the presence of any obstacle minimizes. More specifically, our contributions in this work are summarized below.

– We consider stability maximization for a D2D multicast communication in presence of obstacles and provide a mathematical formulation of it.
– We present a polynomial time algorithm to obtain a maximum stable D2D multicast route.
– Given the knowledge about the blockage probabilities due to the dynamic obstacles, we prove the optimality of our proposed algorithm.
– We then present a technique to estimate these blockage probabilities using the Dempster-Shafer evidential theory framework.
– Finally, through simulations we demonstrate the superiority of our proposed method over a fixed probability distribution based scheme [7] as well as a naïve scheme which is unaware of any obstacles.

The rest of the paper is organized as follows. In Sect. 2 we present the system model along with various assumptions. Mathematical formulations of the considered problem is given in Sect. 3. The proposed stability maximization algorithm is given in Sect. 4 along with the proof of its optimality. The mechanism for leaning the blockage probabilities using evidential theory is presented in Sect. 5. Section 6 demonstrates the benefit of proposed scheme through simulations. Finally, we conclude this work with Sect. 7.

2   System Model

We consider a service area controlled by a central BS. We assume D2D overlay communication scenario in presence of static as well as dynamic obstacles. A multicast group $\mathcal{M}$ consists of a single source UE and n number of other UEs who receive the multicast data from the source UE. We assume all communication to be in half-duplex mode, i.e., a UE can not receive and transmit simultaneously. Time is discretized into slots $t_0, t_1, \cdots$ each with a small Δt time span. We also discretize the entire service area into a grid of small squared cells each of size

$a \times a$ as done in [4, 8]. Furthermore, we consider the UEs to be stationary, i.e. they do not change their positions during the execution of our proposed algorithm and their locations are known to the BS [4].

Static Obstacle Modelling: Similar to [4,8], we assume that the approximate sizes and locations of the static obstacles are known apriori. Such information can be obtained through satellite imagery [8]. A grid cell that contains even a portion of an obstacle is considered to be fully *blocked*. Figure 1 shows the blocked cells (shaded with gray color) due to the presence of an obstacle (black rectangle). As evident, here we

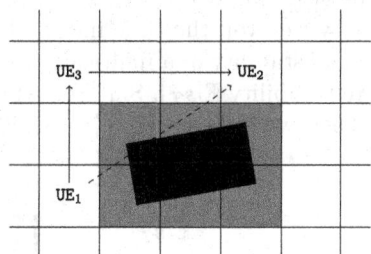

Fig. 1. Communication avoiding an obstacle

essentially overestimate the size of the obstacle. The inaccuracy caused by this overestimation is discussed in Sect. 6.

Modelling Dynamic Obstacles: Suppose the grid cell (x, y) contains a dynamic obstacle with probability $b_{x,y}$. We assume that BS knows the blockage probability $b_{x,y}$ for every grid cell (x, y). We set $b_{x,y} = 1$, whenever there is a static obstacle in the grid cell (x, y). Furthermore, we maintain a matrix called `blockage`, where `blockage`$[x_s, y_s][x_e, y_e]$ denotes the blockage probability of an LOS link between grid cells (x_s, y_s) and (x_e, y_e). Note that this `blockage` matrix incorporates the presence of both static and dynamic obstacles. Thus, if there exist a static obstacle in between two grid cells, the corresponding entry in the `blockage` matrix will be 1. The computation of this probability matrix based on evidential theory is described in Sect. 5.

Multicast Communications: Similar to [10], we assume that a transmitter T_x can simultaneously transmit data to a bunch of receivers R_x residing in close proximity of each other and falls within the beamwidth of the transmitter. Here we consider a multicast group $\mathcal{M}$, where the members of $\mathcal{M}$ are reachable from the source. Although a multicast group member can reside relatively far from the source, it can still receive the packets via other group members, other relay devices [9] or even through the traditional communication via the BS [12].

3 Problem Formulation

Given a set $\mathcal{M}$ of D2D users in a multicast group, let us construct another set $\mathcal{L}$ which consists of possible D2D links $l_{i,j}$, from UE i to UE j, for all $i, j \in \mathcal{M}$. Now let us denote the selection of a link $l_{i,j}$ as $x_{i,j}$. That is,

$$x_{i,j} = \begin{cases} 1 & \text{when } l_{i,j} \text{ is selected in the multicast route} \\ 0 & \text{otherwise} \end{cases}$$

At any given time t, a link $l_{i,j}$ being blocked due the presence of a (dynamic) obstacle is indicated by $b_{i,j}^t = 1$, and 0 otherwise. Let us define the *stability* of a link $l_{i,j}$ at a particular time t as the probability of the link not being blocked by any obstacles and denote it by $s_{i,j}^t$. Note that, $s_{i,j}^t = P\{b_{i,j}^t = 0\}$. For notational simplicity we drop the time index t from the superscripts. Suppose $s_{\mathcal{T}}$ denotes the overall stability of a multicast tree $\mathcal{T}$. Then the objective is to maximize the expected stability $\mathbb{E}[s_{\mathcal{T}}]$. Since the stability of individual links is independent of each other, we have $s_{\mathcal{T}} = \prod_{l_{i,j} \in \mathcal{T}} s_{i,j}$. Thus, we can write

$$\mathbb{E}[s_{\mathcal{T}}] = \mathbb{E}\left[\prod_{l_{i,j} \in \mathcal{T}} s_{i,j}\right] = \prod_{l_{i,j} \in \mathcal{T}} \mathbb{E}[s_{i,j}]$$

The objective function can be given by the linear expression (1).

$$\max \prod_{l_{i,j} \in \mathcal{L}} (x_{i,j} \mathbb{E}[s_{i,j}] + (1 - x_{i,j})) \tag{1}$$

Since the multicast route is essentially a spanning tree, the following constraints are adapted from Martin's formulation [19]. Note that a spanning tree of n nodes contains exactly $n - 1$ edges. This can be encoded as Eq. (2).

$$\sum_{l_{i,j} \in \mathcal{L}} x_{i,j} = |\mathcal{M}| - 1 \tag{2}$$

Next we develop the constraints to ensure there is no cycle in the tree. Thus, it is evident that any UE k can only be connected at only one side of link $l_{i,j}$ whenever $l_{i,j}$ is selected. For this let us first introduce another set of indicator variables $y_{i,j}^k$ and define it as follows.

$$y_{i,j}^k = \begin{cases} 1 & \text{when } l_{i,j} \text{ is selected in the multicast route and} \\ & k \text{ is connected at the side of } j \\ 0 & \text{otherwise} \end{cases}$$

Since a UE k can only be connected at only one side of link $l_{i,j}$ whenever $l_{i,j}$ is selected we have the constraints (3) and (4).

$$y_{i,j}^k + y_{j,i}^k = x_{i,j} \quad \forall l_{i,j} \in \mathcal{L}, \forall k \in \mathcal{M} \tag{3}$$

$$\sum_{k \in \mathcal{M} \setminus \{i,j\}} y_{i,j}^k + x_{i,j} = 1 \quad \forall l_{i,j} \in \mathcal{L} \tag{4}$$

Finally, we have the integrality constraints as given by Eqs. (5) and (6).

$$x_{i,j} \in \{0, 1\} \quad \forall l_{i,j} \in \mathcal{L} \tag{5}$$

$$y_{i,j}^k, y_{j,i}^k \in \{0, 1\} \quad \forall l_{i,j} \in \mathcal{L}, \forall k \in \mathcal{M} \tag{6}$$

Therefore, the maximum stable multicast route can be obtained by solving the integer program where the objective function is given by expression (1) and the constraints are given by Eqs. (2) through (6).

4 Obstacle Aware Multicasting Algorithm

As stated earlier, a multicast route for a particular group is essentially a spanning tree of the UEs in that group [11]. In the following subsection we present an algorithm for constructing a stable multicast route.

4.1 Constructing a Stable Multicast Route

Here the objective is to build a multicast tree that has the maximum possible stability. Recall that the stability $s_{i,j}$ of a particular link $l_{i,j}$ is defined as the probability of not being blocked by any obstacle. Since the stability of two links are uncorrelated, the expected stability $\mathbb{E}[s_T]$ of the entire multicast tree T is essentially the product of expected stability of the individual links in T, i.e., $\mathbb{E}[s_T] = \prod\limits_{l_{i,j} \in T} \mathbb{E}[s_{i,j}]$. The stability of a link $l_{i,j}$ is obtained from the blockage matrix: $s_{i,j} = 1 - \texttt{blockage}[x_i, y_i][x_j, y_j]$. Here (x_i, y_i) and (x_j, y_j) are the grid cells containing the UEs i and j respectively. To obtain such a tree, we first construct a graph $G = (V, E)$, where V has a vertex corresponding to each UE in the multicast group $\mathcal{M}$, and we put an edge between two vertices i and j if their corresponding UE can establish a LOS D2D link between them, i.e., $l_{i,j} \in \mathcal{L}$. Each edge (i, j) in G is given a weight $w_{i,j}$, where $w_{i,j} = -\log_a(\mathbb{E}[s_T])$. Here the base a of the logarithm can be any arbitrary positive constant, and thus can be omitted for notational simplicity. Note that, $\mathcal{L}$ is the set of those links $l_{i,j}$ for which UE i and j are within a distance d_{max} and do not have any static obstacle in between them, i.e., $\texttt{blockage}[x_i, y_i][x_j, y_j] < 1$. Now we construct a minimum spanning tree (MST) T of this graph G using Kruskal's algorithm. We return this T as the required multicast tree. This process formalized as Algorithm 1.

Algorithm 1: Construct Maximum Stable Multicast Tree

```
   /* Populate the set of candidate links L */
1  L = {l_{i,j} | ∀i, j ∈ M s.t. dist(i, j) ≤ d_max and blockage[x_i, y_i][x_j, y_j] < 1}
   /* Construct the graph G = (V, E) */
2  V = {i | ∀i ∈ M}
3  foreach i ∈ V do
4  |   foreach i ∈ V do
5  |   |   if ∃l_{i,j} ∈ L then              // i and j have an LOS D2D link
6  |   |   |_  Put an edge (i, j) in E with weight w_{i,j} = − log(E[s_{i,j}])

7  T ← MST(G)                       // construct the minimum spanning tree
8  return T
```

Here the procedure $\texttt{MST}(G)$ returns a minimum spanning tree of the given graph G. In Algorithm 1, the construction of the graph G takes $O(|\mathcal{M}|^2)$ time. We know that a minimum spanning tree of a graph G can be obtained by Kruskal's algorithm in $O(m \log n)$ time, where n and m are the number of vertices and edges in G respectively. Thus, the step 7 takes $O(|\mathcal{L}| \log |\mathcal{M}|)$ time to execute. Therefore, the overall running time is $O(|\mathcal{M}|^2 + |\mathcal{L}| \log |\mathcal{M}|)$.

Lemma 1. *Algorithm 1 returns a spanning tree having maximum stability.*

Proof. Suppose Algorithm 1 returns $\mathcal{T}^*$. Now consider any other spanning tree T of G for the given multicast group $\mathcal{M}$. Then by definition, we get

$$\sum_{(i,j) \in \mathcal{T}^*} w_{i,j} \leq \sum_{(i,j) \in T} w_{i,j}$$

$$\implies \sum_{(i,j) \in \mathcal{T}^*} -\log(\mathbb{E}[s_{i,j}]) \leq \sum_{(i,j) \in T} -\log(\mathbb{E}[s_{i,j}]) \qquad \text{(by construction)}$$

$$\implies \sum_{(i,j) \in \mathcal{T}^*} \log(\mathbb{E}[s_{i,j}]) \geq \sum_{(i,j) \in T} \log(\mathbb{E}[s_{i,j}]) \qquad \text{(negating both sides)}$$

$$\implies \log \left(\prod_{(i,j) \in \mathcal{T}^*} \mathbb{E}[s_{i,j}] \right) \geq \log \left(\prod_{(i,j) \in T} \mathbb{E}[s_{i,j}] \right) \qquad \text{(taking the log outside)}$$

$$\implies \prod_{(i,j) \in \mathcal{T}^*} \mathbb{E}[s_{i,j}] \geq \prod_{(i,j) \in T} \mathbb{E}[s_{i,j}]$$

$$\text{(since log is an increasing function)}$$

$$\implies \mathbb{E} \left[\prod_{(i,j) \in \mathcal{T}^*} s_{i,j} \right] \geq \mathbb{E} \left[\prod_{(i,j) \in T} s_{i,j} \right]$$

$$\text{(since the random variables are uncorrelated)}$$

$$\implies \mathbb{E}[s_{\mathcal{T}^*}] \geq \mathbb{E}[s_T] \qquad \text{(by definition)}$$

Thus the expected stability of the multicast tree $\mathcal{T}^*$ obtained through Algorithm 1 is as good as that of any other spanning tree T for the given multicast group $\mathcal{M}$. □

5   Learning Blockage Probabilities

A D2D communication link can be blocked by a static as well as by a dynamic obstacle. As stated in Sect. 2, the knowledge of the static obstacles is obtained from satellite imagery. As pointed out by the authors of [8] such information might not be very accurate as the obstacle sizes can be small and may not be captured by the satellite imagery. Whereas, the dynamic obstacles are much harder to deal with, since they move independently and are outside the purview of the BS. There are several methods proposed in the literature to track the movement of these dynamic obstacles, but those solutions requires extra hardware installation and are expensive. Instead, here we present a much cheaper solution to capture the notion of the dynamic as well as the static obstacle. As stated in Sect. 2, we have discretized the service area into small square grids.

Now consider a D2D link between two grid cells (x_1, y_1) and (x_2, y_2). If one can successfully establish a LOS link between (x_1, y_1) and (x_2, y_2), then it is evident that there is no static obstacle on that path. Therefore, we may mark all the grid cells lying on that link to be free from static obstacles. Now if the link is blocked, there might be a static or dynamic obstacle present in some grid cell lying on that path. In this case we mark each grid cell lying on this path between (x_1, y_1) and (x_2, y_2) as potentially containing an obstacle. As more and more links are tried to be established, we gather more such marking information for each grid cells. This information can be combined to obtain an estimate on the blockage probability at each grid cell. Now we need to consider the fact that the channel information itself is not very accurate, as there might be high interference and noise. This presents us with an imperfect knowledge about the environment. Therefore, now a link blockage due to significant drop in SINR value could be due to the presence of an obstacle, or it could be the case that there was a high interference or noise.

To deal with such imperfect knowledge, we can take multiple such measurements and consider the evidences of blockage for a particular grid cell to update our knowledge about the blockage probabilities. We thus consider two type of events for a particular grid cell:

b: there is an obstacle present, the cell should be marked as blocked

f: there is no obstacle, the cell should be marked as free

Furthermore, we consider another case $\{b, f\}$ when we do not know which event has actually occurred. We consider an attempt of a link establishment as a trial. Thus, a trial can present us with one of these three evidences:

$$\{b\}, \{f\} \text{ and } \{b, f\}$$

We now utilize the Dempster-Shafer theory (DST) [20] to deal with such evidences. In our case, the basic probability assignment (bpa) m_{init} is taken as:

$$m_{init} = \{\frac{1}{M}, \frac{M-2}{M}, \frac{1}{M}\}$$

where M is a suitably large value. We initialize evidence masses $m_{x,y}$ of all grid cells (x, y) to m_{init}. Thus we give waitages to $\{b\}$ and $\{b, f\}$ as close to zero while $\{f\}$ is given a remaining waitage which is close to one. This essentially denotes the fact that all grid cells are initially assumed to be free of any obstacles. Whenever a link between (x_1, y_1) and (x_2, y_2) is blocked it provides evidence of blockage for to all grid cells on the line joining (x_1, y_1) and (x_2, y_2). Those grid cells can be efficiently obtained with a subroutine line_points(). More specifically, line_points(i, j) runs the Bresenham line drawing algorithm [21], that uses only integral addition and subtractions, to compute the intermediate grid cells on the line joining the two grid cells containing UE i and UE j. Suppose there are n many grid cells on the line joining (x_1, y_1) and (x_2, y_2), then we formulate the blocking evidence for each of these grid cells as follows:

$$e_b(n) = \{\frac{1}{n}, \frac{1}{M}, \frac{Mn - M - n}{Mn}\}$$

Now we use the Yager's rule [22] to combine this new evidence into the existing knowledge $m_{x,y}$ about any grid cell (x,y) lying on the line joining (x_1, y_1) and (x_2, y_2). Let us denote this combination operation as: $m_{x,y} \leftarrow m_{x,y} \circledast e_b(n)$. Similarly, for an evidence of obstacle free LOS path we update the existing knowledge as: $m_{x,y} \leftarrow m_{x,y} \circledast m_{init}$. Finally, we take the blockage probability $b_{x,y}$ of a grid cell (x,y) as the belief function $\text{bel}_{x,y}(\{b\})$ defined as the probability mass value of $\{b\}$ in $m_{x,y}$. This process is formalized as Algorithm 2.

Algorithm 2: Learning Blockage Probabilities

---

1 **foreach** grid cell (x,y) **do**
2 $\quad \lfloor \; m_{x,y} \leftarrow m_{init}$ // initial mass distribution
3 **while** *true* **do**
4 $\quad \mid \quad \mathcal{L} \leftarrow$ set of links selected by Algorithm 1
5 $\quad \mid \quad \mathcal{B} \leftarrow$ links in $\mathcal{L}$ that could not be activated   // possibly blocked
6 $\quad \mid \quad$ **foreach** link $l_{i,j} \in \mathcal{B}$ **do**
7 $\quad \mid \quad \mid \quad \mathcal{P} \leftarrow \text{line\_points}(i,j)$ // set of grid points between i and j
8 $\quad \mid \quad \mid \quad n \leftarrow |\mathcal{P}|$ // number of grid cells in $\mathcal{P}$
9 $\quad \mid \quad \mid \quad$ **foreach** grid point $(x,y) \in \mathcal{P}$ **do**
10 $\quad \mid \quad \mid \quad \lfloor \; m_{x,y} \leftarrow m_{x,y} \circledast e_b(n)$ // combine the new evidence
11 $\quad \mid \quad$ **foreach** link $l_{i,j} \in \mathcal{L} \setminus \mathcal{B}$ **do**   // obstacle free LOS links
12 $\quad \mid \quad \mid \quad \mathcal{P} \leftarrow \text{line\_points}(i,j)$
13 $\quad \mid \quad \mid \quad$ **foreach** grid point $(x,y) \in \mathcal{P}$ **do**
14 $\quad \mid \quad \mid \quad \lfloor \; m_{x,y} \leftarrow m_{x,y} \circledast m_{init}$ // combine the initial mass
15 $\quad \mid \quad$ **foreach** grid cell (x,y) **do**
16 $\quad \mid \quad \mid \quad b_{x,y} \leftarrow \text{bel}_{x,y}(\{b\})$ // belief of blockage
17 $\quad \mid \quad \lfloor \; s_{x,y} \leftarrow 1 - b_{x,y}$ // stability

---

Note that this is an online learning process that is executed after each time slot and the blockage probabilities are updated accordingly.

Remark 1. In Sect. 6, we demonstrate the convergence of the Algorithm 2 is demonstrated through simulation.

6   Simulation Results

We consider a simulation environment similar to [4,7]. The service area is considered to be of size $1000\,\text{m} \times 1000\,\text{m}$. We take the grid sizes as $5\,\text{m} \times 5\,\text{m}$. The number of UEs in the multicast group $\mathcal{M}$ is varied from 50 to 200. The UEs are placed uniformly at random in the service area. The distance between two D2D pairs is at most $20\,\text{m}$. The static obstacle sizes are considered to be $10\,\text{m} \times 10\,\text{m}$ We consider a fixed number (100) of static obstacles scattered over the service area uniformly at random. Whereas, the sizes of dynamic obstacles are $4\,\text{m} \times 4\,\text{m}$

and very their number from 50 to 200. We assume mmWave D2D communications uses the 60 GHz frequency band. The antennas use beamforming technique and the beamwidth is 45°. The other channel parameters are assumed to be similar as taken in [4].

Figure 2 demonstrates the convergence of our proposed framework for leaning the blockage probabilities obtained through Algorithm 2. We use a Monte-Carlo simulation where we place few dynamic obstacles using Random Walk with maximum velocity 1.5 m/s and measure the link failures. The plot shows a span of a thousand time slots, i.e., a thousand iterations of the outer loop of Algorithm 2. Here we consider the value of bel({b}) for a single grid cell under four different cases where probability of grid being blocked are $0.3, 0.5, 0.7$ and 0.9 respectively. As evident in the figure, the bel({b}) approaches 1 when blocking probability is greater than 0.5. Higher the blocking probability, quicker the value of bel({b}) saturates to 1. In case of blocking probability being lesser than 0.5 the value of bel({b}) stays almost always 0. Whereas if the blocking probability is near about 0.5 the bel({b}) value keeps fluctuating as one would expect.

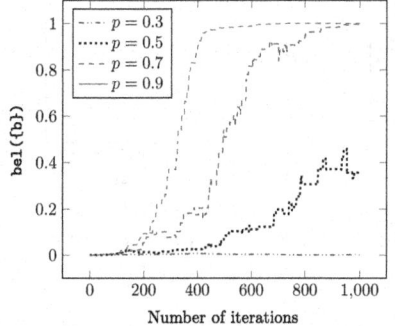

Fig. 2. Convergence of the Learning

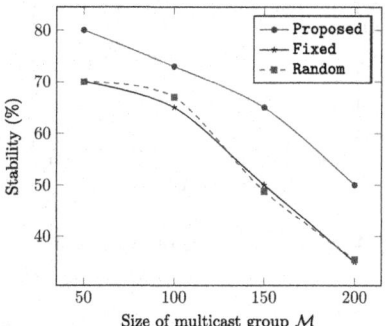

Fig. 3. Stability verses group size

Next We demonstrate the effectiveness of our proposed framework. For this we consider the D2D multicasting application with the objective of maximizing the stability of the route. To obtain a maximum stable multicast tree we use the Algorithm 1 with the learned blockage probabilities. Let us denote this as **Proposed**. To compare the performance of our proposed scheme we consider the approach used in [7]. Here the blockage probabilities are fixed values distributed using a Poisson process. We run the same Algorithm 1 using these probability values. We denote this as **Fixed**. Moreover, we also consider a random spanning tree generation algorithm irrespective of any knowledge of blockage probability. Let us call this algorithm **Random**. We perform a Monte-Carlo simulation where we similarly place few dynamic obstacles following random walks and test how many selected links are affected these obstacles. The stability of a constructed spanning tree is calculated as the percentage of links blocked by the dynamic obstacles. We repeat this process and take an average over that.

Figure 3 plots the stability with varying size of multicast groups with number of dynamic obstacles fixed at 200. As evident in Fig. 3, the `Proposed` scheme perform better than `Fixed` and `Random` scheme. As one would expect, the stability of the constructed multicast tree decreases as the spanning tree grows with the group size. Moreover, Fig. 3 also suggests that using a static fixed distribution does not help much, compared to random selection of links for the spanning tree without considering the presence of any dynamic obstacles. In Fig. 4, we plot the stability with a fixed group size of 200 while we vary the number of dynamic obstacles from 50 to 200. Figure 4 shows that the stability of the constructed multicast tree decreases as the number of obstacles increase, as expected. A similar behavior is observed here also, that is, our `Proposed` scheme performs better than the `Fixed` and `Random` schemes.

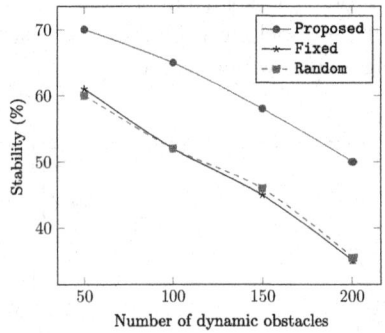
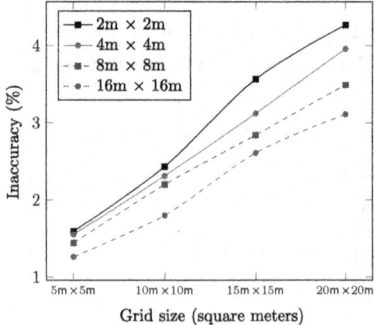

Fig. 4. Stability verses obstacle count   **Fig. 5.** Effect of grid approximation

As mentioned in Sect. 2, we have discretized the service area into grid of small squares. Furthermore, we have also assumed that a grid cell as fully blocked even if it contains only a portion of an obstacle. This approximation overestimates the actual blockage due to the presence of the obstacles. In Fig. 5, we try to capture the effect of this discretization. For this we define a metric called *inaccuracy* which is defined as the ratio $\frac{n-n_a}{n}$ where n is the number of link blockages when we overestimate partial overlap between a grid cell and an obstacle as the grid cell being fully blocked, while n_a is the actual number of link blockages considering exact size and position of the obstacles. We vary the obstacle sizes from $2\,m \times 2\,m$ to $16\,m \times 16\,m$ and also vary the size of grid cells from $5\,m \times 5\,m$ to $20\,m \times 20\,m$ and do a Monte Carlo simulation to measure the inaccuracy. We plot the result in Fig. 5. As one would expect, the inaccuracy increases with e grid cell sizes as more and more overestimate is being done in case of larger grid cells. Moreover, the inaccuracy also increases as the obstacles size reduces as more overestimation is being done, which is also expected.

7   Conclusion

In this work we have addressed the stable route selection problem for multicast D2D communication in presence of static and dynamic obstacles. To this end, we have developed a greedy stable multicast tree construction algorithm. We have proven its optimality assuming the blockage probabilities are known to the BS. Next we have presented a framework for learning these blockage probabilities based on the Dempster-Shafer evidential theory. We have shown the convergence of our proposed framework through simulations. We have also shown the effectiveness of our proposed stable multicast tree construction algorithm over existing approaches through simulations. The choice of size of the grid squares plays a crucial role in the accuracy of proposed framework which captures the blockage probabilities due to the dynamic obstacles. Furthermore, we also present the effect of such grid size choice and overestimation of obstacle sizes through simulation. While finer grid sizes reduces the overestimation due to the grid approximation, it requires more storage space. Thus obtaining a balance between the two could be a possible extension of this work.

References

1. Tehrani, M.N., Uysal, M., Yanikomeroglu, H.: Device-to-device communication in 5g cellular networks: challenges, solutions, and future directions. IEEE Comm. Mag. **52**(5), 86–92 (2014). https://doi.org/10.1109/MCOM.2014.6815897
2. Cai, L.X., Cai, L., Shen, X., Mark, J.W.: Rex: a randomized exclusive region based scheduling scheme for mmWave WPANs with directional antenna. IEEE Trans. Wirel. Commun. **9**(1), 113–121 (2010). https://doi.org/10.1109/TWC.2010.01.070503
3. Bhardwaj, A., Agnihotri, S.: Multicast Protocols for D2D, pp. 1–18. Wiley, Hoboken (2020). https://doi.org/10.1002/9781119471509.w5GRef183
4. Dutta, R.N., Ghosh, S.C.: Resource allocation for millimeter wave D2D communications in presence of static obstacles. In: Barolli, L., Woungang, I., Enokido, T. (eds.) AINA 2021. LNNS, vol. 225, pp. 667–680. Springer, Cham (2021). https://doi.org/10.1007/978-3-030-75100-5_57
5. Hammed, Z.S., Ameen, S.Y., Zeebaree, S.R.M.: Investigation of 5g wireless communication with dust and sand storms. J. Commun. **18**(1), 36–46 (2023). https://doi.org/10.12720/jcm.18.1.36-46
6. Qiao, J., Shen, X.S., Mark, J.W., Shen, Q., He, Y., Lei, L.: Enabling device-to-device communications in millimeter-wave 5G cellular networks. IEEE Comm. Mag. **53**(1), 209–215 (2015). https://doi.org/10.1109/MCOM.2015.7010536
7. Kumar S, Y., Ohtsuki, T.: Influence and mitigation of pedestrian blockage at mmWave cellular networks. IEEE Trans. Veh. Technol. **69**(12), 15442–15457 (2020). https://doi.org/10.1109/TVT.2020.3041660
8. Sarkar, S., Ghosh, S.C.: Relay selection in millimeter wave D2D communications through obstacle learning. Ad Hoc Netw. **114**, 102419 (2021). https://doi.org/10.1016/j.adhoc.2021.102419
9. Chiti, F., Fantacci, R., Pierucci, L.: Social-aware relay selection for cooperative multicast device-to-device communications. Future Internet **9**(4) (2017). https://doi.org/10.3390/fi9040092

10. Ningombam, D.D., Shin, S.: Resource-sharing optimization for multicast D2D communications underlaying LTE-A uplink cellular networks. In: Proceedings of the 34th ACM/SIGAPP Symposium on Applied Computing. SAC '19, pp. 2001–2007. ACM, NY, USA (2019). https://doi.org/10.1145/3297280.3297476
11. Maufer, T.A., Semeria, C.: Introduction to IP Multicast Routing. Internet-Draft draft-ietf-mboned-intro-multicast-03, Internet Engineering Task Force, October 1997. https://datatracker.ietf.org/doc/draft-ietf-mboned-intro-multicast
12. Bettancourt, R.: LTE broadcast multicast. https://futurenetworks.ieee.org/images/files/pdf/applications/LTE-Broadcast-Multicast030518.pdf
13. Li, Y., Kaleem, Z., Chang, K.H.: Interference-aware resource-sharing scheme for multiple D2D group communications underlaying cellular networks. Wirel. Pers. Commun. 90(2), 749–768 (2016). https://doi.org/10.1007/s11277-016-3203-2
14. Gupta, S.K., Khan, J.Y., Ngo, D.T.: A D2D multicast network architecture for vehicular communications. In: 2019 IEEE 89th Vehicular Technology Conference (VTC2019-Spring), pp. 1–6, April 2019. https://doi.org/10.1109/VTCSpring.2019.8746673
15. Huang, J., Zhong, Z., Ding, J.: An adaptive power control scheme for multicast service in green cellular railway communication network. Mob. Netw. Appl. 21(6), 920–929 (2016). https://doi.org/10.1007/s11036-016-0712-x
16. Rubin, I., Tan, C.C., Cohen, R.: Joint scheduling and power control for multicasting in cellular wireless networks. EURASIP J. Wirel. Commun. Netw. 2012, 250 (2012). https://doi.org/10.1186/1687-1499-2012-250
17. Park, J., Heath, R.W.: Analysis of blockage sensing by radars in random cellular networks. IEEE Sig. Process. Lett. 25(11), 1620–1624 (2018). https://doi.org/10.1109/LSP.2018.2869279
18. Koda, Y., Yamamoto, K., Nishio, T., Morikura, M.: Reinforcement learning based predictive handover for pedestrian-aware mmWave networks. In: IEEE INFOCOM 2018 - IEEE Conference on Computer Communications Workshops (INFOCOM WKSHPS), pp. 692–697, April 2018. https://doi.org/10.1109/INFCOMW.2018.8406993
19. Martin, R.: Using separation algorithms to generate mixed integer model reformulations. Oper. Res. Lett. 10(3), 119–128 (1991). https://doi.org/10.1016/0167-6377(91)90028-N
20. Shafer, G.: A Mathematical Theory of Evidence. Princeton University Press, Princeton (1976). https://doi.org/10.1515/9780691214696
21. Bresenham, J.E.: Algorithm for computer control of a digital plotter. IBM Syst. J. 4(1), 25–30 (1965). https://doi.org/10.1147/sj.41.0025
22. Yager, R.R.: On the Dempster-Shafer framework and new combination rules. Inf. Sci. 41(2), 93–137 (1987). https://doi.org/10.1016/0020-0255(87)90007-7

Mobility Aware Path Selection for Millimeterwave 5G Networks in the Presence of Obstacles

Subhojit Sarkar$^{(\boxtimes)}$ and Sasthi C. Ghosh

Advanced Computing and Microelectronics Unit Indian Statistical Institute,
203 B. T. Road, Kolkata 700108, India
subhojitfriends@gmail.com, sasthi@isical.ac.in

Abstract. An often overlooked metric in millimeterwave communication is the so-called stability of assigned links. Links can fail due to obstacles (both static and dynamic), and user mobility. Handling static obstacles is easy; the challenge comes in tracking dynamic obstacles which may impede transmission in the near future. Most works in literature track dynamic obstacles by using additionally deployed hardware like radars and cameras, which add to deployment cost. Our approach requires no such additional hardware. We handle multiple static obstacles, and a single dynamic obstacle, to allocate stable transmission paths which are unlikely to break in the near future. We then adopt a holistic approach to the problem of transmission path allocation, by assigning paths that are estimated to be active for the longest possible time, albeit at the cost of some individual throughput. In other words, we do not always select the best link based on instantaneous system parameters; rather, we can counter-intuitively choose slower, but more stable links. To achieve this, we consider user mobility, and obstacles which may cause links to fail prematurely. We demonstrate the accuracy with which dynamic obstacle trajectories can be captured by our proposed approach. By further simulation studies, we show that our approach leads to the assignment of more stable links as compared to the traditional received signal strength based greedy approach.

Keywords: 5G communication networks · millimeterwaves · ultra dense network · link allocation · obstacle tracking · stability

1 Introduction

Millimeterwave (mmwave) communication in the gigahertz (GHz) range is expected to form the backbone of fifth and sixth generation communication networks [1,2]. This is primarily due to two reasons, namely the exponential rise in demand for high speed communication (especially with technologies like Internet of Things, Vehicle-to-Vehicle Communications, Virtual Reality, and Augmented

© The Author(s), under exclusive license to Springer Nature Switzerland AG 2023
F. Neri et al. (Eds.): CCCE 2023, CCIS 1823, pp. 67–80, 2023.
https://doi.org/10.1007/978-3-031-35299-7_6

Reality fast reaching large scale deployment), and the fast depletion of traditional sub-6GHZ communication channels. It is therefore not surprising this field of study has generated a lot of interest over the past decade [3–5].

Millimeterwaves come with their own share of problems. Their inability to penetrate even thin obstacles [6] lead to their stringent line of sight (LOS) demand during communication. Additionally even under LOS, mmwaves can only provide high speed data transmission over short distances due to their high free space propagation loss. Thus, the main aim of systems deploying mmwave communication is data transmission over short, LOS links. Due to the inherent blockages in the real life transmission environment, the problem of establishment of LOS paths demands novel ideas. Examples include ultra dense networks which deploy a large number of small cell mmwave base stations (BSs), and intelligent reflecting surfaces which augment the transmission environment to bypass obstacles. Additionally, nearby idle user equipment (UE) can volunteer as relays, thereby establishing multi-hop LOS paths between source and destination devices. There are a large number of relay selection algorithms reported in literature, which mostly deal with increasing the instantaneous throughput. Majority of the literature deals with greedy throughput maximization, with less regard for link active time. In other words, the approach is usually to allocate links greedily, to maximise the immediate datarate. Due to user mobility, links are bound to fail over time, sometimes soon after allocation. This leads to repeated link search overheads, decreasing user throughput, and worsening user experience. There has been some works [7,8] that decrease the number of unnecessary handoffs while maintaining quality of service, by smartly assigning mmwave BSs to requesting UEs. However, both have considerable training times, and neither considers the problem of handling dynamic obstacles. Assuming a real life mobility scenario where UEs do not change their velocity and direction too frequently, we propose an alternative approach that allocates transmission paths that have the longest active time. Obviously, the accuracy of the proposed approach depends on the mobility pattern of UEs; if they are too random (an unlikely real life scenario), this approach will provide no improvement in link active time.

We now briefly explain our proposed approach. In Fig. 1a, there are two UEs U_A and U_B which are moving with velocities v_A and v_B respectively in the directions as shown. Their initial positions at time t_0 are indicated by the filled disks; the hollow circles show the estimated positions of the UEs at subsequent time instances. There is a static obstacle near the transmission path linking the two moving UEs. Assuming that the UEs have infrequent change in their velocities and directions, we can fairly estimate the time t_2 at which the link will be broken due to lack of LOS. This becomes our estimated link active time. Similarly, in Fig. 1b, there are two UEs moving parallel to each other, with different velocities as shown. Given the maximum acceptable transmission range t_{max}, we can estimate the time instance t_4, at which the link between U_A and U_B will break due to the distance criterion. The minimum of these two estimates gives us the overall estimate about the link active time. There are many other

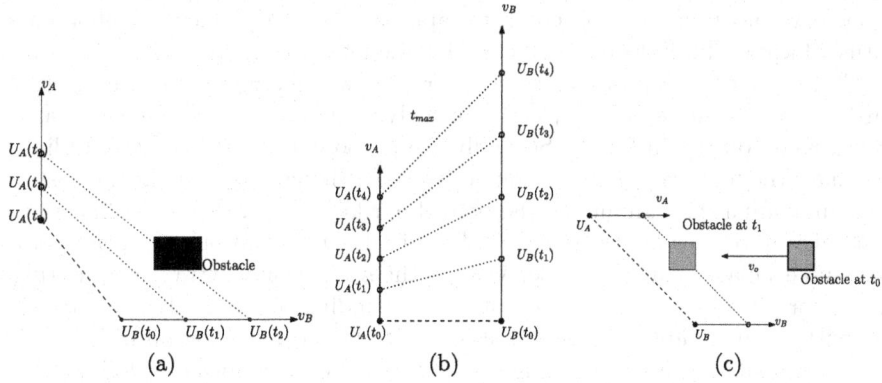

Fig. 1. Link failure due to (a) static obstacle, (b) transmission distance exceeding threshold distance, and (c) dynamic obstacle

cases which may arise (such as UEs travelling in different directions, or one of the devices being a static mmwave BS), which are not shown due to paucity of space. The gist however is thus: assuming that the UEs do not frequently change their velocities and directions, we can allocate links which have higher estimated link active time. Of course there will be some unavoidable link failures; possibly due to UEs entering buildings, leaving the coverage area, sudden change in velocity and direction.

The problem is not confined to static obstacles only. In fact the authors [9] pointed out a high attenuation in mmwave links due to pedestrian blockage. We demonstrate the problem in Fig. 1c, where there are two UEs namely U_A and U_B moving with velocities v_A and v_B respectively in the direction as shown. There is also a dynamic obstacle moving with velocity v_o in the direction as shown. As is evident, the dynamic obstacle will cause link failure in the near future, more specifically at time t_1. The usual way to deal with this problem is to deploy additional cameras at the BSs [10] or radars [11] throughout the coverage area, that track and estimate trajectories of dynamic obstacles. More recently LiDARs [12,13] have been used for blockage tracking, and subsequent fast switching from mmwave to sub-6GHZ even before the LOS is broken. The authors [14] proposed an approach to predictively handover soon-to-be-broken links by following a single pedestrian, using a human tracking module. They then subsequently used reinforcement learning (RL) to extract mobility patterns, and decide on an optimal timing to perform predictive handover. These approaches solve the problem, but with significant additional cost overhead. The cost associated with such large scale deployment of expensive equipment will undoubtedly be of concern to network providers, and consequently will financially burden the end users. In this paper, we use existing hardware to track a single dynamic obstacle, and allocate links that lie spatially outside its estimated trajectory, thereby reducing the chance of link failure. We consider a single dynamic obstacle as in [14], and leverage existing well-known systems to track its estimated trajectory, and propose a probabilistic approach to the problem by allocating paths which are likely to remain active for the longest time period.

The main contributions of this paper are twofold; firstly we use existing hardware to estimate the trajectory of a dynamic obstacle which may cause transmission outage in future. Secondly, we propose a method to allocate links that have the highest estimated link active time, though arguably at the cost of some individual throughput. To the best of our knowledge, this is the first work that takes into consideration the mobility of UEs, the locations of static obstacles, and even a dynamic obstacle, without the need for any extra hardware (like radars, or vision cameras). The problem of handling multiple dynamic obstacles without any additional hardware, seems to be a challenging problem and is left as a possible future work. The rest of this paper is arranged as follows. We describe our assumed system model, and the proposed algorithm in Sects. 2, and 3 respectively. We validate the proposed approach in Sect. 4 using simulation, concluding in Sect. 5.

2   System Model

In this section, we briefly describe the considered system model.

Coverage Area: The coverage area under consideration is discretized into small square grids, whose resolution can be limited only by hardware constraints. However, high resolutions come with their fair share of problems, including costly time and space requirements. A very high resolution would have increased the efficiency of our approach. However, we considering the real life equipment limitations, we consider a moderately small resolution.

System Architecture: We consider an ultra dense [15], urban deployment scenario. A central long term evolution (LTE) BS provides ubiquitous coverage. Our path allocation algorithm runs at this central BS. There are $\mathcal{F} = \{F_1, F_2, \cdots, F_f\}$ femtocells (mmwave BSs) distributed uniformly at random inside the coverage area. Each such femtocell is connected to the central LTE BS, as well as to each other by a high speed backhaul network. There are sufficient number of channels at each mmwave BS to serve the requests that may come. In other words, the domain of spatial reuse is outside the domain of this paper. In case no mmwave BS can serve a request due to absence of short, LOS connectivity, the LTE BS steps in and provides sub-6GHz service.

UE and UE Mobility: There is a set $\mathcal{U}$ of mmwave enabled UEs moving about inside the coverage area. These can be pedestrians, cyclists, or users in vehicles equipped with mmwave devices. One group $\mathcal{D}$ of these UEs demands a high-speed link to any mmwave BS, while the remaining UEs $\mathcal{R}$, are idle and can serve as relays for other nearby UEs that lack a 1-hop LOS path to any mmwave BS. In the coverage area, a UE moves in a manner resembling the random waypoint mobility model. It selects direction and velocity uniformly at random from a given range, and continues to travel for a specific time epoch. The process repeats after one such epoch. In real life, UEs do not change their velocity or direction very frequently. Indeed, intersections in roadways form a small percentage of the total road network. In other words, we can predict the future location of a UE, to a good accuracy, based on the past trajectory. We use this intuition to allocate links which will not break soon. All UEs have electronically steerable directional antennas [16] and mmWave transreceivers, to compensate for the high attenuation of mmWaves. Each UE has a superaccurate GPS chip [17], an accelerometer, and a compass. As a part of the location update process, yhe UE communicates its location and velocity to the LTE BS. We assume that a relaying UE relay the data of a single demanding UE.

Obstacle Modelling: There is a set $\mathcal{S} = \{S_1, S_2, \cdots, S_s\}$ of static obstacles distributed uniformly at random inside. For tractability, we assume that all the obstacles are axes parallel rectangles, an assumption that can easily be removed in practice. The positions and dimensions of these obstacles are known beforehand, which can be done from satellite imagery [18], or by using a learning method [19]. We consider that no additional construction takes place after deployment; if such a scenario occurs, it can be easily incorporated in our algorithm by updating $\mathcal{S}$. There is a single dynamic obstacle moving inside the coverage area, following a similar mobility pattern as the UEs. This can be a pedestrian, or a vehicle like a car or bus. Unlike the UEs, this obstacle is not connected to any of the BSs. As such, the LTE BS has no information regarding the position, velocity, and past trajectory of the said obstacle.

Throughput Calculation: Due to the fast attenuating nature of mmwaves, we assume that the maximum transmission distance under LOS conditions is t_{max}; above this threshold distance, the attenuation is so high, that the usage of mmwave bands is not justified. Furthermore, we do not consider transmission under non-LOS conditions in this paper. If there is no path from a UE in $\mathcal{U}$ to any mmwave BS even via a relay, transmission has to take place over sub-6GHz band via the LTE BS.

The path loss between two nodes i and j, $PL(d_{i,j})$ is calculated as in [20]:

$$PL(d_{i,j}) = \alpha + \beta 10 log_{10}(d_{i,j}) + \zeta. \tag{1}$$

Here, α and β are the pathloss parameters, and ζ is a log-normal random variable with zero mean, and $d_{i,j}$ is the Euclidean distance between the two nodes. We use $y \longleftrightarrow z$ to denote that the devices y and z can communicate with each other via mmwaves.

3 Proposed Path Allocation Algorithm

We divide up the algorithm into two parts, one dealing with the dynamic obstacle, and the other dealing with the UE mobility and static obstacles.

3.1   Estimating Dynamic Obstacle Trajectory

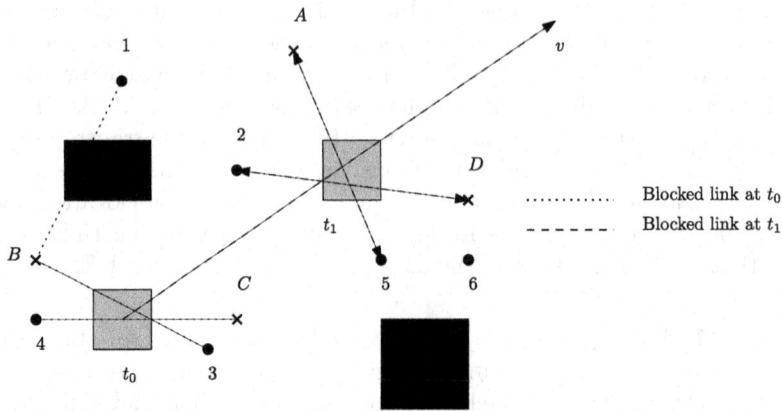

Fig. 2. Estimating Possible Trajectories

Our idea uses a modified version of the signal space partitioning scheme proposed in [8]. In [8], the authors estimated the spatial location of a UE by using a notion of signal partitioning, without resorting to pinpointing the exact geographical location. Our idea involves both signal partitioning and accurate geographical locations to get an idea regarding the presence of a dynamic obstacle in a transmission path. In the toy example shown in Fig. 2, the crosses ($\times$) enumerated with letters are the mmwave BSs, and the disks ($\bullet$) enumerated with numbers are the UEs. The dark black squares represent the known static obstacles, and the light grey squares represent the positions of a single dynamic obstacles at time slots t_0 and t_1. The dynamic obstacle is moving in the shown direction with a velocity v. The LTE BS has prior knowledge regarding the locations and dimensions of static obstacles; it also knows the locations of the UEs. As a result for the location of a UE r, the LTE BS is aware of L_r, the list of all mmwave BSs from which it is supposed to receive signals (under the absence

Table 1. Signal Partitioning

UE (r)	L_r	$L'_r(t_0)$	$L'_r(t_1)$
1	A	A	A
2	A, B, C, D	A, B, C, D	A, B, C
3	B, C	C	B, C
4	B, C	B	B, C
5	A, C, D	A, C, D	C, D
6	C, D	C, D	C, D

of any dynamic obstacle). At a time t, a UE receives signals from all mmwave BSs that are within its close LOS range, and sends the list $L'_r(t)$ to the LTE BS. Comparing this list with L_r, we get an idea regarding the presence of a dynamic obstacle. For simplicity, we assume that the UEs are static between the time t_0 to t_1 (the idea can be very easily applied to dynamic UEs). In Fig. 2, we see that links $3 - B$ and $4 - C$ are blocked at time t_0. This tells us there is a dynamic obstacle on their spatial intersection position (as we have considered a single dynamic obstacle). We note that at time t_0, even link $1 - B$ is blocked; however, since the location of the static obstacle is known beforehand, we do not take this as a marker for dynamic obstacle. Similarly at time t_1, we find that links $2 - D$ and $5 - A$ are blocked, which tells us the updated location of the dynamic obstacle. We summarize this in Table 1.

Using the two intersecting locations and the time interval $(t_1\text{-}t_0)$, we estimate the trajectory of the dynamic obstacle. We point out here that it may very well happen that the dynamic obstacle will not obstruct any link at multiple time instances, thereby making position extraction impossible. However for ultra dense networks, this happens in rare cases, as validated in the simulation section.

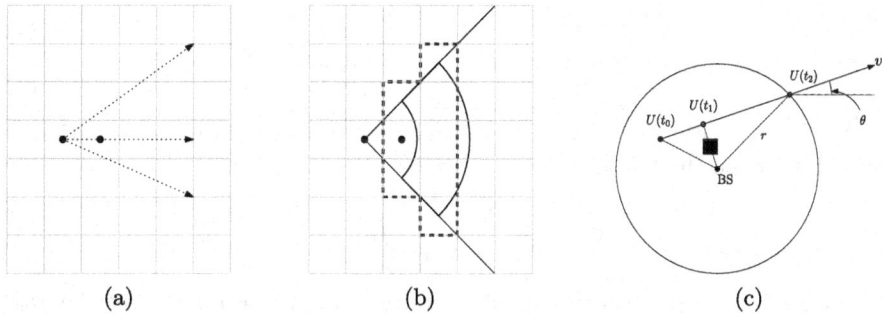

Fig. 3. (a) Possible trajectory cone of dynamic obstacle, (b) Possible zone of obstruction due to dynamic obstacle, and (c) Estimating possible zone of obstruction by dynamic obstacle

Remark I: There is however, a subtle additional point that has to be taken care of. The finite grid resolution introduces inaccuracy in the predicted path. In other words, trajectories at different angles may very well have same intersecting grids. In Fig. 3a, the positions of the dynamic obstacle at time t_0 and t_1 are shown by disks. As is evident, the finite grid resolution maps multiple possible trajectories to the same points. The larger the resolution, the larger is the estimated *trajectory cone*. To deal with this as shown in Fig. 3b, we enumerate the maximal set of possible trajectories of the dynamic obstacle. If we have more and more data points regarding the positions of the dynamic obstacle, this cone shrinks. In other words, the larger the difference between two time slots, the greater the accuracy; however, too long a time interval increases the chance of the dynamic obstacle changing its velocity, thereby leading to inaccurate trajectory prediction.

Remark II: We point out here that our approach cannot be applied very easily for multiple obstacle scenario. This is because in such a case, there is no way of deducing which of the obstacles obstructed the transmission (or indeed, if multiple of them did). For situations that involve multiple dynamic obstacles, we can learn the traffic mobility patterns [19] and allocate paths avoiding zones of congestion. We leave the problem of handling multiple dynamic obstacles without using any additional hardware as our future work.

3.2   UE Mobility and Static Obstacles

We now handle the static obstacles, and UE mobility, In Fig. 3c, there is a single static mmwave BS at (b_x, b_y), a static obstacle at position (o_x, o_y), and a mobile UE. At time t_0, the location of the UE is at $U(t_0) = (x_0, y_0)$, velocity is v, and it is moving at an angle θ with the horizontal. From high school geometry, the position of the UE at time t will be

$$x_t = x_0 + v \times \cos(\theta) \times t \tag{2}$$

$$y_t = y_0 + v \times \sin(\theta) \times t \tag{3}$$

According to our mobility pattern, we can find the time varying equation of the link between the UE and the BS as:

$$y - y_t = \frac{b_y - y_t}{b_x - x_t} \times (x - x_t) \tag{4}$$

Plugging in the position of the static obstacle (o_x, o_y) in Eq. 4, we get the estimated time t_1 at which the link will be broken due to the static obstacle lying on its transmission path. Note that an additional checking has to be done to ensure that the obstacle lies on the line segment joining the two devices, and not merely on the extended line. Also, it might very well be the case that t_1 is negative; it simply implies that in such a case, the said obstacle will never block the link under consideration as long as the velocity is unchanged. A similar checking has to be done to estimate t_2, the time for which the UE will be within the maximum transmission distance t_{max}, from the BS. That is,

$$\sqrt{(x_t - b_x)^2 + (y_t - b_y)^2} \leq r \tag{5}$$

Plugging Eqs. (2) and (3) in Eq. (5) we get a second degree equation of t which gives us t_2. For non-zero v we can get one of the following scenarios:

- two imaginary values, indicating that said UE is never within transmission range of the BS

Algorithm 1: Trajectory cone estimation of a dynamic obstacle

Input: $L_\mathcal{R}$, $L'_\mathcal{R}[t_1, \cdots, t_W]$
Output: $\vec{C}$, $\vec{v}$

1 **for** $i \leftarrow 1 : W$ **do**
2   **for** $r \in \mathcal{R}$ **do**
3         $Y_r^i \leftarrow L_r \oplus L'_r(t_i)$
4   **for** $(r_1, r_2) \in \mathcal{R}^2 \mid r_1 \neq r_2$ **do**
5   **if** $Y_{r_1}^i \vee Y_{r_2}^i \neq 0$ **then**
6             $\mathbb{T}^i \leftarrow intersect(Y_{r_1}^i, Y_{r_2}^i, r_1, r_2)$
7   break

8 Using $\mathbb{T}$ and W, estimate $\vec{C}$ and $\vec{v}$
9 **return** $\vec{C}$, $\vec{v}$

- two unequal real values, indicating UE will lie inside the transmission range of the BS within this time interval. In this case we take the least positive value as t_2, the time after which the UE is likely to move out of the transmission range. In case both values are negative, we can safely ignore the considered obstacle.
- two equal real values, indicating the UE is moving tangentially to the coverage area of the BS

The final estimated transmission time for the link (due to static obstacles and UE mobility) is the minimum of the two values t_1 and t_2. The UE-UE case can be handled exactly similarly.

3.3 Proposed Algorithm

Armed with the locations of static obstacles $\mathcal{S}$, and the possible trajectories of the dynamic obstacle, we can now formally describe the proposed algorithms. In the pre-processing stage the LTE BS computes the set of all transmissible mmwave BSs from each grid location. This is done by incorporating the static obstacles $\mathcal{S}$, and the maximum transmission range t_{max}. The following array based implementation is done to generate L, with i being a grid position and $j \in \mathcal{F}$.

$$L(i, j) = \begin{cases} 1, & \text{if } i \longleftrightarrow j \\ 0, & \text{otherwise} \end{cases} \tag{6}$$

After deployment, each node in $\mathcal{R}$ sends the list of mmwave BSs from which it is receiving signals, to the LTE BS. Using this data over a specific time window W, the LTE BS generates $L'_\mathcal{R}(t)$. For a given window W, $L_\mathcal{R}$ and $L'_\mathcal{R}(t)$ become the input of Algorithm 1, $L_\mathcal{R}$ being the subset of L corresponding to the grid locations of $\mathcal{R}$. We calculate the exclusive OR (X-OR) of the two input arrays for each time slot i and each relay r, and store it in Y_r^i. This gives us an efficient

---

Algorithm 2: Greedy Path Allocation Algorithm

Input: $\mathcal{D}$, $\mathcal{R}$, $\mathcal{F}$, $\mathcal{S}$, t_{max}, $\overrightarrow{C}$, $\overrightarrow{v}$
Output: $\mathbb{P}$

1 **for** $i \in \mathcal{D} \bigcup \mathcal{R}$ **do**
2   **for** $j \in \mathcal{R} \bigcup \mathcal{F} \mid i \neq j$ **and** $dynamic(i, j, \overrightarrow{C}, \overrightarrow{v})$ **do**
3        $T_{ij}^1 \leftarrow obstacle(i, j, \mathcal{S})$
4        $T_{ij}^2 \leftarrow transmit(i, j, t_{max})$
5        $T_{est} \leftarrow min(T_{ij}^1, T_{ij}^2)$
6   **if** $T_{est} > 0$ **then**
7           $addEdge(\mathbb{G}, i, j, T_{est})$

8 **for** $f \in \mathcal{F}$ **do**
9     $addEdge(\mathbb{G}, f, \mathbb{F}, INF)$
10 **for** $d \in \mathcal{D}$ **do**
11    $P_d \leftarrow widestPath(\mathbb{G}, d, \mathbb{F})$
12   **if** $P_d \neq \phi$ **then**
13   Add P_d to $\mathbb{P}$
14   Remove non-BS nodes in P_d from $\mathbb{G}$
15   **else**
16   Serve d via sub-6GHz band

17 **return** $\mathbb{P}$

---

measure of those links that are blocked due to a dynamic obstacle. For a relay pair (r_1, r_2) at time i, the logical OR of $Y_{r_1}^i$ and $Y_{r_2}^i$ gives us the spatial intersection points of the two links, and is subsequently used as a point in the trajectory $\mathbb{T}$ of the dynamic obstacle in the $intersect()$ function. Repeating this step over the window W, we get a set of points along with the corresponding time slots. Using $\mathbb{T}$ and W, we estimate trajectory cone $\overrightarrow{C}$ and velocity $\overrightarrow{v}$ of the dynamic obstacle. The computation takes into account the effect of grid resolution as remarked in Section 4.1.

Using $\overrightarrow{C}$, $\overrightarrow{v}$, and the locations and velocities of all the UEs, Algorithm 2 greedily allocates paths to all demanding users in $\mathcal{D}$. We create a visibility graph $\mathbb{G} = (V, E)$ where the nodes are communicating devices (all UEs and all mmwave BSs). There is an edge between two nodes only if (a) the link does not lie in $\overrightarrow{C}$, and (b) the estimated link active time between the two devices is strictly positive. The $dynamic()$ function estimates if the link ij will be obstructed by the dynamic obstacle. The $obstacle()$ function estimates the time to failure of a link due to a static obstacle in $\mathcal{S}$, while the $transmit()$ function estimates the time for which a link will be active due to the maximum transmission range criterion. The weight of an edge in $\mathbb{G}$ is the minimum of the two estimates.

We run a widest path algorithm *widestPath*() on $\mathbb{G}$, starting from each $d \in \mathcal{D}$, and terminating at the supernode $\mathbb{F} = \bigcup_{\mathcal{F}} f$. This is a modified version of Dijkstra's shortest path algorithm as suggested in [21]. It returns P_d, the path beginning at d and ending at any $F \in \mathcal{F}$, and having the longest estimated active time. If the *widestPath*() function finds a path, the same is added to $\mathbb{P}$; the participating non-BS nodes are removed from $\mathbb{G}$. In case P_d is empty, data transmission takes place over sub-6GHz service.

4   Simulation Results

The modelling parameters are mostly adapted from [7]. The path loss parameters are $\alpha = 61.4$ and $\beta = 2$, and the thermal noise density is -174 dBm/Hz. We consider a square area of size $100 \times 100 \ m^2$ as the coverage area under the LTE BS, the grid resolution being $1m$. An LTE BS provides ubiquitous coverage, while some mmwave BSs provide high speed, short range services. There are users with mmwave enabled devices moving around inside with speeds ranging from [0,10m/s]; 50% of the UEs require a high speed link to any mmwave BS, while the rest are willing to act as relays. There are known static obstacles inside the coverage area, size of each being 5 m × 5 m. There is a single dynamic obstacle moving around following the same model as a UE.

We plot the dynamic obstacle tracking accuracy of our algorithm in Fig. 4. Our approach would not have been possible in sparsely deployed networks, with the trajectory being identified in only 20% of the cases for 10 mmwave BSs and 50 relay UEs. However, as we move towards ultra dense networks and heavy user density, we see that the trajectory of the dynamic obstacle can be obtained accurately in upto 90% of the cases. We now define *average link active time*, T_{active} as the average time to failure of all allocated paths, which may occur due to either obstacles, or UE mobility. In Fig. 5, we plot the effect of the maximum velocity (V_{max}) of UEs on the average link active time. The maximum velocity is varied from 5 m/s to 30 m/s. We see that for low speeds, the proposed approach outperforms the traditional RSS based approach by a significant amount. As V_{max} increases, the difference becomes smaller, although the proposed method continues to outperform the RSS based method. In Fig. 6, we plot the effect of the mmwave base station density on the average link active time. The mmwave BS count is varied from 5 to 30. We see that for a 100 m × 100 m area, the average link active time stabilizes around BS count of 20, and the proposed algorithm continues to provide more stable links than the RSS based approach. The effect of the number of static obstacles inside the coverage area on the average link active time is shown in Fig. 7. The obstacle count is varied from 2 to 10; since each obstacle is of size $25 \ m^2$, the number of grids covered is obtained by multiplying the obstacle count by 25. As is obvious, with increasing number of static obstacles, the average link active time falls for both the methods, though the proposed method continues to outperform the traditional method.

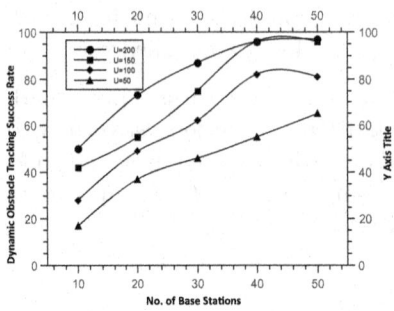

Fig. 4. Obstacle tracking accuracy

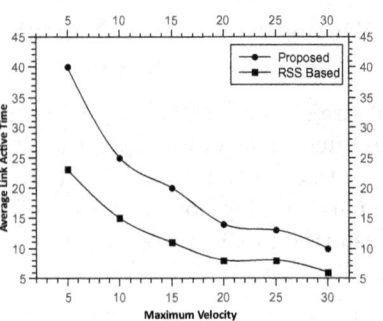

Fig. 5. Max. UE velocity vs T_{active}

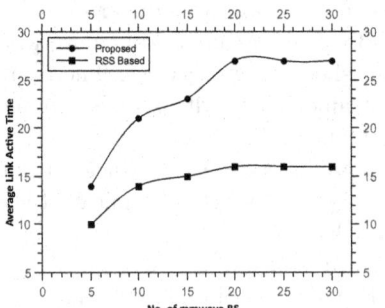

Fig. 6. No. of mmwave BS vs T_{active}

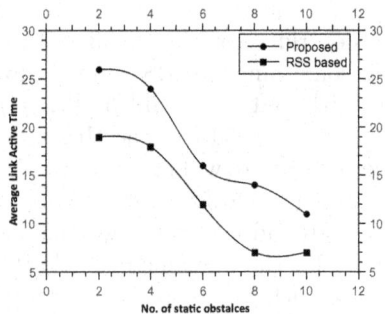

Fig. 7. No. of static obstacles vs T_{active}

5   Conclusion

In this paper, we propose a simple algorithm to track a dynamic obstacle inside a transmission environment without using any additional hardware. We subsequently allocate mmwave transmission paths which have the longest estimated time to failure, by taking into consideration static obstacles, UE mobility, and the trajectory of the dynamic obstacle. We show via simulation that our approach provides higher average link active times than the usual RSS-based approach. Some of the obvious future directions of this work can be handling fairness index, inter-UE interference, and spatial reuse in resource constrained scenarios. In future, we will also tackle the more challenging problem of path allocation while avoiding multiple dynamic obstacles, without resorting to any additional hardware.

References

1. Rappaport, T.S., et al.: Millimeter wave mobile communications for 5G cellular: it will work! IEEE Access **1**, 335–349 (2013)

2. Sandi, E., Rusmono, R., Diamah, A., Vinda, K.: Ultra-wideband microstrip array antenna for 5g millimeter-wave applications. J. Commun. **15**(2), 198–204 (2020). https://doi.org/10.12720/jcm.15.2.198-204

3. Wang, X., et al.: Millimeter wave communication: a comprehensive survey. IEEE Commun. Surv. Tutor. **20**(3), 1616–1653 (2018). https://doi.org/10.1109/COMST. 2018.2844322

4. Busari, S.A., Huq, K.M.S., Mumtaz, S., Dai, L., Rodriguez, J.: Millimeter-wave massive MIMO communication for future wireless systems: a survey. IEEE Commun. Surv. Tutor. **20**(2), 836–869 (2018)

5. Kutty, S., Sen, D.: Beamforming for millimeter wave communications: an inclusive survey. IEEE Commun. Surv. Tutor. **18**(2), 949–973 (2016)

6. Pi, Z., Khan, F.: An introduction to millimeter-wave mobile broadband systems. IEEE Commun. Mag. **49**(6), 101–107 (2011)

7. Sun, Y., Feng, G., Qin, S., Liang, Y.C., Yum, T.S.P.: The smart handoff policy for millimeter wave heterogeneous cellular networks. IEEE Trans. Mob. Comput. **17**(6), 1456–1468 (2018). https://doi.org/10.1109/TMC.2017.2762668

8. Sun, L., Hou, J., Shu, T.: Spatial and temporal contextual multi-armed bandit handovers in ultra-dense mmWave cellular networks. IEEE Trans. Mob. Comput. **20**(12), 3423–3438 (2021). https://doi.org/10.1109/TMC.2020.3000189

9. Koda, Y., Nakashima, K., Yamamoto, K., Nishio, T., Morikura, M.: Handover management for mmWave networks with proactive performance prediction using camera images and deep reinforcement learning. IEEE Trans. Cogn. Commun. Netw. **6**(2), 802–816 (2020)

10. Alrabeiah, M., Hredzak, A., Alkhateeb, A.: Millimeter wave base stations with cameras: vision-aided beam and blockage prediction. IEEE Vehicular Technology Conference (2020)

11. Park, J., Heath, R.W.: Analysis of blockage sensing by radars in random cellular networks. IEEE Signal Process. Lett. **25**(11), 1620–1624 (2018)

12. Zheng, Y., Chen, S., Zhao, R.: A deep learning-based mmWave beam selection framework by using lidar data. In: 2021 33rd Chinese Control and Decision Conference (CCDC), pp. 915–920 (2021)

13. Zhang, T., Liu, J., Gao, F.: Vision aided beam tracking and frequency handoff for mmWave communications. In: IEEE INFOCOM 2022 - IEEE Conference on Computer Communications Workshops (INFOCOM WKSHPS), pp. 1–2 (2022)

14. Koda, Y., Yamamoto, K., Nishio, T., Morikura, M.: Reinforcement learning based predictive handover for pedestrian-aware mmWave networks. In: IEEE INFOCOM - IEEE Conference on Computer Communications Workshops, pp. 692–697 (2018)

15. Baldemair, R., et al.: Ultra-dense networks in millimeter-wave frequencies. IEEE Commun. Mag. **53**(1), 202–208 (2015)

16. Liu, J., Kato, N., Ma, J., Kadowaki, N.: Device-to-device communication in LTE-advanced networks: a survey. IEEE Commun. Surv. Tutor. **17**(4), 1923–1940 (Fourthquarter 2015). https://doi.org/10.1109/COMST.2014.2375934

17. Moore, S.K.: Super-accurate GPS coming to smartphones in 2018 [news]. IEEE Spectr. **54**(11), 10–11 (2017). https://doi.org/10.1109/MSPEC.2017.8093787

18. Hu, Q., Blough, D.M.: Relay selection and scheduling for millimeter wave backhaul in Urban environments. In: 2017 IEEE 14th International Conference on Mobile Ad Hoc and Sensor Systems (MASS), pp. 206–214 (2017). https://doi.org/10.1109/ MASS.2017.48

19. Sarkar, S., Ghosh, S.C.: Relay selection in millimeter wave D2D communications through obstacle learning. Ad Hoc Netw. **114**, 102419 (2021). https://doi.org/10. 1016/j.adhoc.2021.102419

20. Akdeniz, M.R., et al.: Millimeter wave channel modeling and cellular capacity evaluation. IEEE J. Sel. Areas Commun. **32**(6), 1164–1179 (2014). https://doi.org/10.1109/JSAC.2014.2328154
21. Pollack, M.: Letter to the editor-the maximum capacity through a network. Oper. Res. **8**(5), 733–736 (1960). https://doi.org/10.1287/opre.8.5.733

A Probabilistic Analysis of the Delay in RIS Assisted SISO D2D Communication Using Chernoff's Bounds

Durgesh Singh[1] and Sasthi C. Ghosh[2(✉)]

[1] School of Computer Science, University of Petroleum and Energy Studies, Dehradun 248007, India
[2] Advanced Computing and Microelectronics Unit, Indian Statistical Institute, Kolkata 700108, India
sasthi@isical.ac.in, mobile.wifi@gmail.com

Abstract. Reflecting intelligent surface (RIS) assisted millimeter wave (mmWave) device to device (D2D) communication in single input single output (SISO) mode is considered. RIS can bypass the blocked D2D pairs since mmWaves are highly susceptible to blockages. However, the RIS is composed of passive elements which cannot accurately measure the state of the given channel and may suffer outages from moving obstacles (blockages) or due to moving devices (communication out of range) causing uncertainty in link quality. Each data packet will have some probability to be successfully transmitted towards the destination device via a RIS in each time unit. When a packet is lost it has to be re-transmitted which will induce extra delay. In this paper, the impact of link success probability on delay is studied. A lower bound on delay such that all packets are received successfully for the given link success probability has been derived. Later, we have derived a new term called *apparent* success probability which captures the trade-off between delay and link success probability. Simulations are performed which conforms with the derived theoretical results.

Keywords: Reflecting Intelligent Surface (RIS) · D2D communication · Obstacles · Chernoff's bound · Link Success Probability

1 Introduction

There is an exponential surge in data rate requirements in recent past and is expected to increase rapidly for the upcoming future too. Device to device (D2D) communication along with millimeter wave (mmWave) is an apt choice for 5G and beyond technologies for short distance communication. D2D communication enables a device or user equipment (UE) to communicate among each other either directly or via other UEs acting as relays with or without the assistance of a base station (BS) [2,13,17,20–22]. Millimeter waves are the electromagnetic waves between 30-300 GHz spectrum which are studied extensively due to high available bandwidth for D2D communication. D2D communication along with mmWave

F. Neri et al. (Eds.): CCCE 2023, CCIS 1823, pp. 81–92, 2023.
https://doi.org/10.1007/978-3-031-35299-7_7

technology have potential to provide the requirement of very high data rate for short distances. However, mmWave suffers from very severe propagation and penetration losses [19]. The problem for propagation loss can be easily solved, however the penetration loss is still a challenge especially due to motion of UEs and presence of moving obstacles. Recently, reflecting intelligent surface (RIS) has been studied to bypass obstacles and mitigate their effects of blockages in between a given D2D pair, thereby increasing the overall system throughput [1,9,23,26–28]. An RIS is an array of reflecting elements where each element can independently reflect the incoming signal towards the direction of the destination UE. Each element is a PIN diode which has the capability to induce phase shift on the incident wave to reflect it to the desired direction where destination UE resides. Along with increasing the system throughput without any extra transmit power, RIS are advantageous due to small size, light weight and hence convenient to deploy and thereby incurring low overall cost [28,29]. Hence these can be installed on high buildings, trees, poles and high rise structures in a region where UEs frequently face outages due to out of range communication or blockages.

The RIS is used to bypass communication between D2D pair ij when the direct communication channel between ij is facing outages. However, there are following two issues while taking the assistance from the RIS. First, the link between RIS and a UE is also susceptible to blockages/outages due to obstacles and UE's motion. Second, the elements of the RIS are passive devices, thus lacking the facility of transmitting or receiving pilot signals [9]. Hence getting the channel state information in a time efficient manner is very challenging in RIS assisted D2D communication. These issues make the quality of link between a UE and RIS uncertain. For a given fixed RIS installation, the link quality can be enhanced by either installing some expensive technology alongside the installed RIS or by changing the dynamics of the environment where RIS is installed. Both of these solutions require extra cost in some form and thus are difficult to achieve in practice. We intend to enhance the overall link success probability at the cost of inducing extra delay incurred in sending multiple copies of a given packet. The use case under consideration is assumed to have a negligible impact due to extra delay on overall quality of experience by UEs. This requires a probabilistic analysis to capture the performance of RIS links especially the delay induced due to uncertain link qualities for a given RIS assisted D2D pair.

In this paper, we are seeking the solution to the problem of finding overall delay for RIS links for the single input single output (SISO) scenarios such that all the intended packets are received at the receiver with very high probability. A single packet can be transferred successfully from source UE to destination UE in a given time unit with probability p using assistance from RIS. The probability of link success/failure is known from the past observations. A given message is assumed to be of length m packets. Time is discretized as $t, t+1, \cdots$ such that at each time unit a single packet can only be transferred. In a SISO communication, this problem can be viewed as the number of time units n required to send m packets to the destination with very high probability $(1 - \epsilon)$ given that the channel has success probability of p independently for each time unit. We are utilizing Chernoff bounds to give a bound on the number of packets (and hence the required time units) that need to be sent on a single channel in SISO scenario such that exactly m packets are received by the receiving UE. Chernoff bounds give the exponentially decreasing bounds on the tail distribution of failure probability. Later, using the expression for p, we derive an apparent link success probability p'. Apparent link success probability is not the actual change in the probability of blockage but it is the virtual change in p on the cost of varying delay. We argue that by introducing very small delay in order to send multiple copies of few randomly chosen packets, we can enhance the link success probability in terms of apparent probability p'. We have performed simulations to verify that analytical results conform with that of the simulation results.

2   Related Works

A tutorial and a survey for RIS assisted communication are provided by the authors in [9,23] respectively. There are many challenges and opportunities associated using RIS assisted communication systems [27,28]. The authors in [1] considered the performance evaluation of RIS assisted vehicular communication system with multiple non-identical interferers. The authors in [8,24], and [7] discussed about improving the overall sum-rate problem. The RIS assisted mmWave D2D communication is being performed by the authors in [6,16], [10], which is beneficial due to its short range communication. It has been studied for SISO communication systems by the authors in [5,11,12] to solve the localization problem. Due to passive nature of RIS, there are chances of errors to be induced while computing the link quality. The outage probability has been discussed by the authors in [25]. There have been numerous studies performed by several authors in identifying the size and number of RIS elements required to provide required data rate. In [29], the author gave bounds on number of reflecting elements for guaranteed energy- and spectral-efficient RIS assisted communication system. The authors in [15] studied the relationship between the number of reflective elements and system sum rate in RIS assisted communication system. The goal of the work in [14] is to minimize the number of reflecting elements needed depending upon the blockage scenario and power constraints. However, to the best of our knowledge, the impact on extra delay incurred due to outages has not been considered. In this paper, we have performed the analysis for the RIS assisted mmWave D2D communication with SISO scenario. In future, we would leverage this work to more complex scenarios.

3   System Model

We are considering a service region where there are M RIS installations. Each RIS is equipped with N independent reflecting elements which can support at most N different independent communication channels. Each element can reflect the incident signal to some desired direction by suitably changing the phase shift of the incident signal. Each such RIS installation is serving some D2D pair ij comprising of source UE i and destination UE j communicating in SISO scenario. In this case, both source and destination UEs have single antennas for transmission and reception, however, multiple passive elements can be used. For the RIS assisted D2D SISO scenario one of the reflecting elements of the RIS is used to reflect the incident signal of the source UE i towards the destination UE j for the given pair ij. Time is discretized as t, $t+1$, $\cdots$ and Δt is the small time difference between two consecutive time units t and $t+1$. At each time unit t, a single packet is transmitted on a given channel from i to j via RIS on a two hop communication. It is assumed that Δt duration is enough to transmit only one packet from i to j via RIS on a two hop communication. We are assuming that UEs take part in D2D communication using out-band or in-band overlay scenarios such that they are not interfered from the cellular users [18]. Devices are moving independently of each other and the links among them are formed independently of each other. Hence without loss of generality, we can perform analysis for a given D2D pair and a given RIS installation. The analysis will remain valid for all other D2D pairs and their respective serving RIS installations.

The service region is occupied with a number of moving obstacles whose information is not known a priori, which may cause blockage to the UE-RIS or RIS-UE links which may in turn cause outage. The UE motion may also be the reason for such link outage. Hence, we assume that for a given RIS installation, p is the cumulative success probability of the link from source UE i to the destination UE j via the RIS, which can be obtained statistically from past observations. Since for a given RIS installation, the reflecting elements can independently reflect incident signal and they all are in the same spatial location at the RIS installations, the link success probability can be assumed identical and independent for each reflecting element.

The channel coefficients between the source UE i (transmitter) and the given RIS is $\mathbf{g}_i \in \mathbb{C}^{N \times 1}$. Similarly, the channel coefficient between RIS and the destination UE j (receiver) is $\mathbf{h}_j \in \mathbb{C}^{N \times 1}$. We assume that they obey the Rayleigh distribution. The diagonal phase shift matrix of the RIS is $\Theta = diag(e^{j\theta_1}, e^{j\theta_2}, \cdots, e^{j\theta_N})$, where $\theta_k, \forall k \in \{1, 2, \cdots, N\}$ represents the phase shift of kth element [23]. Assuming that there is no interference from other devices on the receiving UE j, the signal y_j received on j can be written as:

$$y_j = \sqrt{a_i} \mathbf{h}_j^H \Theta \mathbf{g}_i x_i + \nu_j \tag{1}$$

where x_i and a_i are the transmitted signal and the transmitted power of the source UE i respectively and ν_j is the additive white Gaussian noise with zero mean and σ^2 variance at the receiver j. Now the signal to noise (SNR) ratio at UE j can be written as follows:

$$S_j = \frac{a_i |\mathbf{h}_j^H \Theta \mathbf{g}_i|^2}{\sigma^2} \tag{2}$$

A link ij is formed between transmitting UE i and receiving UE j via the RIS if the SNR at j is above a required SNR threshold S_γ. The maximum achievable rate of the RIS assisted SISO link can be given by $\nabla = \log_2(1 + S_j)$ [23].

4   Problem Formulation

We are given an RIS installation where each of its elements can be selected independently for every source destination pair for reflecting the source UE's signal towards the direction of the destination UE where the direct source destination path is facing outages. Total number of m packets need to be sent from source to the destination UE. Note that at least m rounds or time units will be needed to transmit all such m packets on a SISO link assuming perfect channel conditions. The objective is to find out the number of packets (including the retransmitted ones) to be sent such that all m packets are received at the destination with very high probability on the given channel considering the link success/failure probability. This will in turn be used to compute the total time units (end to end delay) required to ensure that all m packets have been delivered successfully with very high probability. If we need a total of n packets to be sent to ensure that the receiver has received all the m packets, then we can say that in n time units all packets will be sent with very high probability $(1 - \epsilon)$. Here, $\epsilon \in (0, 1)$ and it can take an arbitrarily small value. Since each packets can be sent independently at each time unit, we can write the event of sending n packets (denoted as X) as follows:

$$X = \sum_{t=1}^{n} X_t \tag{3}$$

Here X_t is an indicator random variable denoting that at time t, a packet is sent successfully on the given channel $(X_t = 1)$, or a packet loss occurred $(X_t = 0)$. Overall n packets are required to ensure that m packets are received at the receiver where link success probability is p. It is evident that X_t follows Binomial distribution.

Let us denote μ as the mathematical expectation of X (i.e. $E[x]$). Now we can write μ as follows:

$$\mu = n \times p \tag{4}$$

Now we can formally state our objective is to find a value of n such that all m packets are sent with very high probability $1 - \epsilon$ given p. That is,

$$P(X \geq c_1 m + c_2) \geq 1 - \epsilon \tag{5}$$

for some constants c_1 and c_2 which are derived in the next section. We also give the analysis for computing a bound on n and some useful insights.

5   A Bound on Delay (n) for RIS Assisted D2D SISO Link

In this section we will first derive a value for n and then prove that the derived value is a lower bound on it such that exactly m packets are received. Chernoff bounds are useful in giving an exponentially decreasing bounds on the tail distributions of a random variable. Now we will state the Chernoff bound in following remark, which will be used to derive the bound on n.

Remark 1. For a random variable $Y = \sum_{l=1}^{n} Y_l$, where all Y_l are independent Binomial trials such that the expectation is $Y_\mu = E[Y]$, Then using Chernoff's bound, we can write for $\delta \in (0, 1)$,

$$P(X \leq (1 - \delta)Y_\mu) \leq e^{-Y_\mu \delta^2/2}. \tag{6}$$

We have to show that $P(X \geq c_1 m + c_2) \geq 1 - \epsilon$. We can also say that μ is a function of both m and p since n is a function of m. Now we can re-write our objective in Eq. (5),

$$P(X \geq \mu p) \geq 1 - \epsilon \tag{7}$$

assuming $c_1 m + c_2 = \mu p$. Now, letting $X = n$ in above inequality, we get,

$$n \geq \frac{c_1 m + c_2}{p^2}.$$

Substituting $c_1 = 1$ and $c_2 = 2\frac{\ln 1/\epsilon}{p}$ in above equation we get,

$$n \geq \frac{m}{p^2} + \frac{2\ln 1/\epsilon}{p^3}. \tag{8}$$

Proposition 1. *The number of packets $\frac{m}{p^2} + \frac{2\ln 1/\epsilon}{p^3}$ as derived in Eq. (8) is a lower bound to n such that exactly m packets are received with very high probability $1 - \epsilon$.*

Proof. We need to show that $P(X \geq \mu p) \geq 1 - \epsilon$ from Eq. (7). From the derived value of $n = \frac{m}{p^2} + \frac{2\ln 1/\epsilon}{p^3}$ from Eq. (8), we have

$$P(X \geq np^2) \geq 1 - \epsilon$$

$$P(X \geq \left\{\frac{m}{p^2} + \frac{2\ln 1/\epsilon}{p^3}\right\}p^2) \geq 1 - \epsilon$$

$$P(X \geq m + \frac{2\ln 1/\epsilon}{p}) \geq 1 - \epsilon \tag{9}$$

We can also re-write Eq. (7) as follows,

$$P(X \leq \mu(1 - p)) \leq \epsilon. \tag{10}$$

Now we need to show the above Eq. (10) is true. Since $p \in (0, 1)$, using Remark 1, we get,

$$P(X < (1 - p)\mu) \leq e^{-\mu p^2/2}. \tag{11}$$

Substituting $\mu = np = \frac{m}{p} + 2\frac{\ln(1/\epsilon)}{p^2}$ in Eq. (11), we get,

$$P(X < (1 - p)\mu) \leq e^{-\left\{\frac{m}{p} + 2\frac{\ln(1/\epsilon)}{p^2}\right\}\frac{p^2}{2}} \tag{12}$$

$$= e^{-\left\{mp + \ln(1/\epsilon)\right\}} \tag{13}$$

$$< e^{-\ln(1/\epsilon)} \tag{14}$$

$$= \epsilon \tag{15}$$

Hence we have proved our proposition.

It can be seen from the proposition that the time units n required to send all m packets is inversely proportional to p. When the value of p increases, value of n decreases, signifying that when success probability of the link is higher, fewer time or rounds are needed. The lowest value which n can take is m for perfect channel condition. We can see as $p \to 1$, $n \to m$ and when $p \to 0$, $n \to \infty$. When link success probability of the given link is 0, then there is infinite time required to send the message on that link which is very intuitive. Hence this relation shows the trade-off between the link success probability and the number of time units (delay) required to send all the data packets successfully to the destination. We will utilize this insight from Proposition 1 into our final result in corollary mentioned below which introduces the new term apparent link success probability.

5.1   Apparent Link Success Probability p'

For a given fixed RIS installation, the link success probability p is also fixed. We intend to derive apparent link success probability p' which is the virtual change in p on the cost of varying delay due to sending duplicate packets randomly. For some randomly chosen packets, we send its multiple copies which will introduce a small delay. This is required to ensure that atleast one packet is sent successfully. Before deriving the apparent success probability we derived the expression of actual link success probability using above Proposition 1 in the following corollary.

Corollary 1. *The actual link success probability can be simplified as:*

$$p = \sqrt[3]{\frac{m}{n} + 2\frac{\ln 1/\epsilon}{n}} \tag{16}$$

Proof. Using the Eq. (8) and relaxing the criteria that exactly $m/p \geq m$ packets are received by the receiver successfully with probability $1 - \epsilon$, we can derive n as follows.

$$n \geq \frac{m}{p^3} + 2\frac{\ln 1/\epsilon}{p^3} \tag{17}$$

Now using above Eq. (17), we can derive a bound on actual link success probability p as follows,

$$p \geq \sqrt[3]{\frac{m}{n} + 2\frac{\ln 1/\epsilon}{n}} \tag{18}$$

If n' time units are used to send overall multiple copies of packets and a total of $n'' \leq n'$ are received successfully, then we can derive $p' > p$ as follows.

Proposition 2. *For the case when, $\frac{n''n}{n'} - m > 2\ln(1/\epsilon)$, the achieved apparent success probability of a given RIS assisted D2D link is more than the actual success probability of the given link.*

Proof. It is easy to see that after incorporating extra delay n', we get an expression for the new apparent link success probability p' using Corollary 1 as follows,

$$p' \geq \sqrt[3]{\frac{m+n''}{n+n'} + 2\frac{\ln 1/\epsilon}{n+n'}} \qquad (19)$$

We can simply solve for the condition $p' > p$ and can get the required result.

Here we have computed the apparent success probability p' by substituting the value of extra delay which is incurred while exploring the quality of the given RIS assisted D2D link. There are two cases which can arise here, scenario (i) when all multiple copies packets are always successful and scenario (ii) when few of the copies of a packet are lost. For the first scenario, $n'' = n'$. It can be easily seen that $p' > p$ which signifies that the apparent link success probability is more than that of the actual link success probability. This is due to the fact that extra cost in terms of delay due to multiple copies of a packet has been paid to increase the overall apparent link success probability. Note that p' is not the actual change in the probability of blockage but it is the virtual change in p on the cost of varying n (delay). Similarly for the second scenario, $n'' \leq n'$, we can get the required result using the Proposition 2. This approach can help us to design a communication system with better apparent blockage probability by tuning n' in a controllable way according to the need of an application.

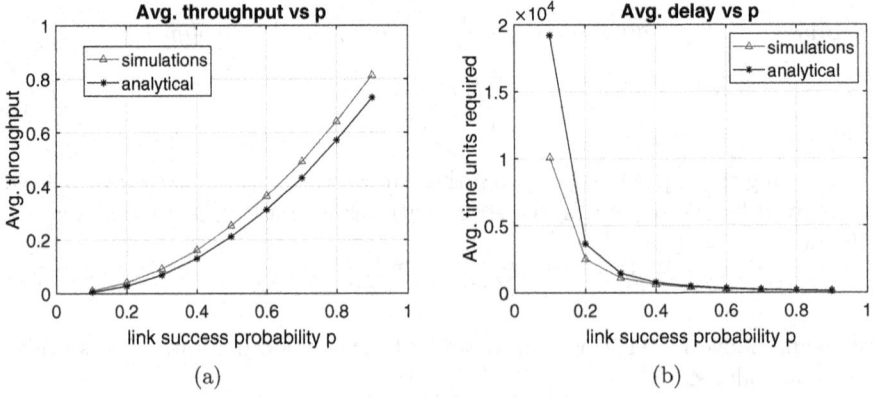

Fig. 1. (a) Average Throughput versus p. (b) Average time units n required to send all m packets versus p.

6   Simulation and Results

We have adapted the simulation environment from the work presented in [3,4] to generate the channel realizations considering 3D Euclidean space. We have

considered the outdoor environment where the RIS is installed on the opposite wall on the YZ plane and we are ignoring angle of arrival and departures.
The authors have provided the Matlab code[1] for channel coefficient matrix for
transmitter-RIS and RIS-receiver channels which is used here for generation of
the same. For simplicity, we have assumed that there is one transmitter receiver
D2D pair (positioned respectively at $(0, 7, 0)$ and $(1, 12, 0)$) whose direct path is
facing outages and they are communicating with assistance from the RIS positioned at $(2, 10, 2)$ where each coordinate is in meters. Transmission power is
assumed to be 23 dBm. We have assumed line of sight (LOS) communication in
transmitter-RIS and RIS-receiver channels. The frequency is 28 GHz, LOS path
loss exponent is 1.98 and noise power is -100 dBm. Shadowing is assumed to be
normally distributed with zero mean and σ^2 variance with σ assumed to be 3.1.
Number of elements N of RIS is 64. We have generated 1000 different channel
realizations. Using the channel coefficient matrix we have written our own $C++$
program to compute throughput and total time required to send entire message
by taking average over these 1000 different channel realizations. The throughput
is computed as ratio of total packets received successfully to that of total packets
transmitted. The overall delay in time units is equal to total number of packets
re-transmitted.

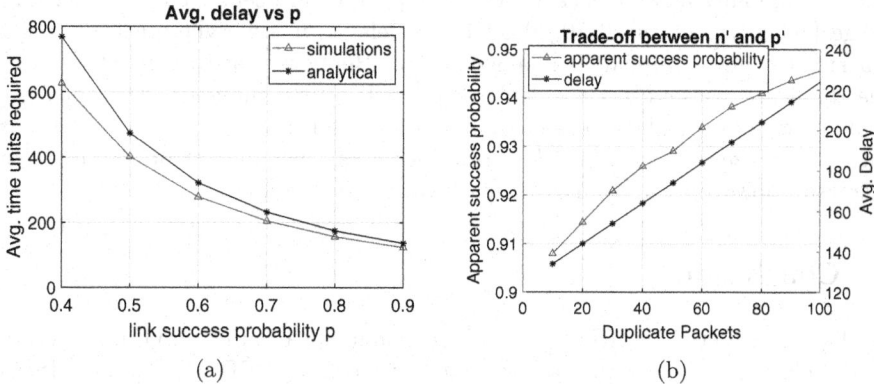

Fig. 2. (a) Zoomed results of Fig. 1(b) for time units n required to send all m packets
versus p. (b) Trade-off between n' and p'.

There are total of $m = 100$ packets need to be sent to the receiver, in case
a packet is lost, that packet is re-transmitted. One time duration Δt is enough
to send a single packet from transmitter to receiver. Considering that there are
dynamic obstacles present in the environment and they may abruptly block the
channels. To successfully send a packet both transmitter-RIS and RIS-receiver
channels must be free from presence of obstacles. Here, p is link success probability of a given channel. We have assumed ϵ to be $1/m$. If n packets are required to

---

[1] Available at https://corelab.ku.edu.tr/tools/simris/.

be sent so that m packets are received successfully, then analytical throughput is m/n and total time units required are n.

Figure 1(a) shows the effect of varying probability of success of a link on throughput. We can see that as the probability of success increases, throughput also increases. This is due to the fact that as probability of success of a link increases, packet delivery chances also increases. The figure also shows the analytical results along with the simulation result. The gap between them is very narrow and is due to the simplification of the current simulation environment.

Figure 1(b) shows the effect of varying probability of success of a link on end to end delay. We can see that as the probability of success increases, total packets (including re-transmission) or total time required to sent all packets is decreasing. This is due to the fact that as probability of success of a link increases, packet delivery chances also increases and hence lesser number of re-transmissions which will in turn use fewer time slots to successfully transmit all data. Here also, the gap between analytical results and the simulation result is very narrow and is due to the simplification of the current simulation environment. Figure 2(a) shows the zoomed in result of Fig. 1(b) for $p \geq 0.4$.

Figure 2(b) shows the trade-off of varying n' on the apparent link success probability p' and the overall delay. For this result, we have kept $p = 0.9$ and keeping all other parameters same as aforementioned. The extra delay n' is due to randomly chosen packets whose duplicates are sent. Here, n' varies in range $\{10, 20, 30, 40, 50, 60, 70, 80, 90, 100\}$. Delay is $n + n'$, where n is obtained via simulations as already explained above. Here, we have assumed that all the duplicate packets are received (scenario (i)) for simplicity. The apparent probability p' is calculated using simulations which is total packets received over total packets sent. We can see that as n' increases, the apparent probability also increases, however, the total average delay is also increased.

7   Conclusion

We have considered the RIS assisted D2D communication in SISO mode where D2D links face outages due to obstacles or motion of UEs. Also, the RIS to UE link qualities are uncertain due to passive beamforming. Hence their is an uncertainty associated with n which is the amount of time units required in order to send all packets to the destination UE. Using Chernoff's bound we have computed the lower bound on n such that exactly m packets are received with very high probability $1 - \epsilon$. The derived expression for n is inversely proportional to p which is intuitive and using this expression we have derived apparent success probability p' of a link. The apparent success probability of a link captures the trade-off between delay and reliability of a message. Through simulations we have numerically shown the impact of derived results on throughput which conforms with the derived theoretical results.

References

1. Agrawal, N., Bansal, A., Singh, K., Li, C.P.: Performance evaluation of RIS-assisted UAV-enabled vehicular communication system with multiple non-identical interferers. IEEE Trans. Intell. Transp. Syst. **23**(7), 9883–9894 (2022). https://doi.org/10.1109/TITS.2021.3123072

2. Ahmed, R.E.: A low-overhead multi-hop routing protocol for D2D communications in 5G. J. Commun. **16**(5), 191–197 (2021)

3. Basar, E., Yildirim, I.: SimRIS channel simulator for reconfigurable intelligent surface-empowered mmWave communication systems, pp. 1–6 (2020)

4. Basar, E., Yildirim, I., Kilinc, F.: Indoor and outdoor physical channel modeling and efficient positioning for reconfigurable intelligent surfaces in mmWave bands. IEEE Trans. Commun. **69**, 8600–8611 (2021)

5. Bhushan, S., Suggala, S.T., Kherani, A.A., Sreejith, S.T.: Approximations and their validations for RIS-assisted SISO systems. In: 2021 IEEE International Conference on Advanced Networks and Telecommunications Systems (ANTS), pp. 452–455 (2021). https://doi.org/10.1109/ANTS52808.2021.9936976

6. Deb, S., Ghosh, S.C.: An RIS deployment strategy to overcome static obstacles in millimeter wave D2D communication. In: 2021 IEEE 20th International Symposium on Network Computing and Applications (NCA), pp. 1–8 (2021). https://doi.org/10.1109/NCA53618.2021.9685506

7. Deb, S., Ghosh, S.K., Ghosh, S.C.: A multi-arm-bandit based resource block allocation in RIS assisted wireless network. In: 2021 IEEE 20th International Symposium on Network Computing and Applications (NCA), pp. 1–6 (2021). https://doi.org/10.1109/NCA53618.2021.9685708

8. ElMossallamy, M.A., Zhang, H., Song, L., Seddik, K.G., Han, Z., Li, G.Y.: Reconfigurable intelligent surfaces for wireless communications: principles, challenges, and opportunities. IEEE Trans. Cogn. Commun. Netw. **6**(3), 990–1002 (2020). https://doi.org/10.1109/TCCN.2020.2992604

9. Gong, S., et al.: Toward smart wireless communications via intelligent reflecting surfaces: a contemporary survey. IEEE Commun. Surv. Tutor. **22**(4), 2283–2314 (2020). https://doi.org/10.1109/COMST.2020.3004197

10. Jia, S., Yuan, X., Liang, Y.C.: Reconfigurable intelligent surfaces for energy efficiency in D2d communication network. IEEE Wirel. Commun. Lett. **10**(3), 683–687 (2021). https://doi.org/10.1109/LWC.2020.3046358

11. Keykhosravi, K., Keskin, M.F., Seco-Granados, G., Popovski, P., Wymeersch, H.: RIS-enabled SISO localization under user mobility and spatial-wideband effects. IEEE J. Select. Topics Sig. Process. **16**(5), 1125–1140 (2022). https://doi.org/10.1109/JSTSP.2022.3175036

12. Keykhosravi, K., Keskin, M.F., Seco-Granados, G., Wymeersch, H.: SISO RIS-enabled joint 3D downlink localization and synchronization. In: ICC 2021 - IEEE International Conference on Communications, pp. 1–6 (2021). https://doi.org/10.1109/ICC42927.2021.9500281

13. Kim, J., Kang, S.: Spectrum allocation with power control in LBS based D2d cellular mobile networks. J. Commun. **15**(3), 283–288 (2020)

14. Li, D.: Bound analysis of number configuration for reflecting elements in IRS-assisted D2D communications. IEEE Wirel. Commun. Lett. **11**(10), 2220–2224 (2022). https://doi.org/10.1109/LWC.2022.3197614

15. Li, D.: How many reflecting elements are needed for energy- and spectral-efficient intelligent reflecting surface-assisted communication. IEEE Trans. Commun. **70**(2), 1320–1331 (2022). https://doi.org/10.1109/TCOMM.2021.3128544

16. Peng, Z., Liu, X., Pan, C., Li, L., Wang, J.: Multi-pair D2D communications aided by an active RIS over spatially correlated channels with phase noise. IEEE Wirel. Commun. Lett. **11**(10), 2090–2094 (2022). https://doi.org/10.1109/LWC.2022.3193868

17. Salhani, M., Liinaharja, M.: Load balancing in UDN networks by migration mechanism with respect to the D2D communications and E2E delay. J. Commun. **14**(4), 249–260 (2019)

18. Singh, D.: Efficient relay selection techniques for D2D communication under user mobility and presence of obstacles, Ph. D. thesis, Indian Statistical Institute, Kolkata (2021)

19. Singh, D., Chattopadhyay, A., Ghosh, S.C.: To continue transmission or to explore relays: Millimeter wave D2D communication in presence of dynamic obstacles. IEEE Transactions on Mobile Computing. p. 1 (2022). https://doi.org/10.1109/TMC.2022.3160764

20. Singh, D., Ghosh, S.C.: Mobility-aware relay selection in 5G D2D communication using stochastic model. IEEE Trans. Veh. Technol. **68**(3), 2837–2849 (2019)

21. Song, X., Han, X., Xu, S.: Joint power control and channel assignment in D2D communication system. J. Commun. **14**(5), 349–355 (2019)

22. Tehrani, M.N., Uysal, M., Yanikomeroglu, H.: Device-to-device communication in 5g cellular networks: challenges, solutions, and future directions. IEEE Commun. Mag. **52**(5), 86–92 (2014)

23. Wu, Q., Zhang, S., Zheng, B., You, C., Zhang, R.: Intelligent reflecting surface-aided wireless communications: a tutorial. IEEE Trans. Commun. **69**(5), 3313–3351 (2021). https://doi.org/10.1109/TCOMM.2021.3051897

24. Xu, K., Zhang, J., Yang, X., Ma, S., Yang, G.: On the sum-rate of RIS-assisted MIMO multiple-access channels over spatially correlated Rician fading. IEEE Trans. Commun. **69**(12), 8228–8241 (2021). https://doi.org/10.1109/TCOMM.2021.3111022

25. Yang, L., Yang, Y., da Costa, D.B., Trigui, I.: Outage probability and capacity scaling law of multiple ris-aided networks. IEEE Wirel. Commun. Lett. **10**(2), 256–260 (2021). https://doi.org/10.1109/LWC.2020.3026712

26. Yildirim, I., Basar, E.: Channel modelling in RIS-empowered wireless communications, pp. 123–148 (2023). https://doi.org/10.1002/9781119875284.ch7

27. Yuan, X., Zhang, Y.J.A., Shi, Y., Yan, W., Liu, H.: Reconfigurable-intelligent-surface empowered wireless communications: challenges and opportunities. IEEE Wirel. Commun. **28**(2), 136–143 (2021). https://doi.org/10.1109/MWC.001.2000256

28. Zhang, H., Di, B., Bian, K., Han, Z., Poor, H.V., Song, L.: Toward ubiquitous sensing and localization with reconfigurable intelligent surfaces. Proc. IEEE **110**(9), 1401–1422 (2022). https://doi.org/10.1109/JPROC.2022.3169771

29. Zhang, H., Di, B., Han, Z., Poor, H.V., Song, L.: Reconfigurable intelligent surface assisted multi-user communications: how many reflective elements do we need? IEEE Wirel. Commun. Lett. **10**(5), 1098–1102 (2021). https://doi.org/10.1109/LWC.2021.3058637

System Security Estimation
and Analysis of Data Network

Enhancing IoT Security Through Deep Learning-Based Intrusion Detection

A. Jyotsna$^{(\boxtimes)}$ and E. A. Mary Anita

Department of Computer Science and Engineering, Christ (Deemed to Be University), Bangalore, India
jyoterance@gmail.com

Abstract. The Internet of Things (IoT) has revolutionized the way we interact with technology by connecting everyday devices to the internet. However, this increased connectivity also poses new security challenges, as IoT devices are often vulnerable to intrusion and malicious attacks. In this paper, we propose a deep learning-based intrusion detection system for enhancing IoT security. The proposed work has been experimented on IoT-23 dataset taken from Zenodo. The proposed work has been tested with 10 machine learning classifiers and two deep learning models without feature selection and with feature selection. From the results it can be inferred that the proposed work performs well with feature selection and in deep learning model named as Gated Recurrent Units (GRU) and the GRU is tested with various optimizers namely Follow-the-Regularized-Leader (Ftrl), Adaptive Delta (Adadelta), Adaptive Gradient Algorithm (Adagrad), Root Mean Squared Propagation (RmsProp), Stochastic Gradient Descent (SGD), Nesterov-Accelerated Adaptive Moment Estimation (Nadam), Adaptive Moment Estimation (Adam). Each evaluation is done with the consideration of highest performance metric with low running time.

Keywords: IoT-23 · Intrusion Detection · Edge Computing · Gated Recurrent Units · Adaptive Moment Estimation

1 Introduction

The Internet of Things (IoT) refers to the growing network of physical devices, such as sensors, smartphones, cameras, and wearable devices, that are connected to the internet and capable of collecting and exchanging data [1]. These devices can be used in various fields including healthcare, transportation, and smart cities [2]. Intrusion detection in IoT devices is the process of identifying unauthorized or malicious activity on these devices and networks [3], and is an important aspect of IoT security to detect and prevent cyber-attacks. Intrusion detection systems for IoT can be categorized as signature-based or anomaly-based, depending on their method of detecting known attacks or deviations from normal behavior [4].

Intrusion detection is a crucial aspect of computer network security as it allows organizations to identify and respond to unauthorized access attempts [5]. The increasing

F. Neri et al. (Eds.): CCCE 2023, CCIS 1823, pp. 95–105, 2023.
https://doi.org/10.1007/978-3-031-35299-7_8

volume and complexity of network traffic, as well as new types of attacks, have made intrusion detection a challenging task [6]. Traditional methods such as rule-based and signature-based systems have been shown to be inadequate in dealing with these challenges. Deep learning, a subfield of machine learning, has been gaining attention in the field of intrusion detection as it can learn from large amounts of data and automatically extract features, making it well-suited for detecting complex and unknown attacks [7]. Research studies have proposed using deep learning for intrusion detection in various scenarios, including network-based, host-based, and web application intrusion detection [8].

Deep learning intrusion detection uses neural networks to evaluate network traffic data for intrusion patterns. A feedforward neural network trained on labeled network traffic data can classify network traffic as normal or abnormal. The input to the network is the network traffic and the output is a label indicating whether the traffic is normal or anomalous. Once trained, the network can be used to classify new, unseen network traffic as normal or anomalous [9].

Another approach is to use recurrent neural networks (RNN) to detect intrusions in sequences of network traffic data. RNNs are well-suited for this task because they can process sequential data and maintain a state that can identify patterns over time. The RNN is trained on labeled sequence data, where the input to the network is a sequence of network packets and the output is a label indicating whether the sequence contains an intrusion or not [10].

Another deep learning architecture that has been used for intrusion detection is convolutional neural networks (CNN). They are particularly well-suited for identifying patterns in spatial data such as images and time-series data like network traffic data. In this approach, the CNN is trained on labeled network traffic data, where the input to the network is a set of network packets and the output is a label indicating whether the set contains an intrusion or not [11].

Adversarial networks have also been proposed as a deep learning approach for intrusion detection. In this approach, a generative model is trained to generate normal network traffic, and a discriminative model is trained to identify whether a given network traffic is normal or anomalous. The generative model is used to generate synthetic data that is used to train the discriminative model. This approach is useful in situations where labeled data is scarce. Reinforcement learning can also be used to train deep neural networks for intrusion detection. In this approach, a neural network is trained to take actions that maximize a reward signal, such as correctly identifying an intrusion [12].

Overall, deep learning approaches for intrusion detection have shown to be effective in identifying complex and unknown attacks, however, the interpretability and explainability of the models remains a challenge.

Edge ecosystems consist of IoT devices, networks that connect them, and the data they produce. These ecosystems can make use of a distributed framework for security by processing data at the source or nearby devices to identify anomalies and protect against potential threats [13]. This often requires communication between multiple edge devices. Common types of security attacks in edge ecosystems include DDoS, malicious code injection, eavesdropping, and tampering. However, traditional security countermeasures used in cloud security cannot be applied in edge ecosystems due to resource constraints, the need for multiple administration, and the requirement of dispersed collaboration between edge nodes. It's challenging to identify anomalies in an edge ecosystem as opposed to a cloud ecosystem due to its decentralized form of computation. However, intrusion detection at the edge should be designed to use minimal resources to ensure the system can detect attacks while not overburdening edge nodes [14].

2 Related Works

The majority of the studies that are included below focused on classification frameworks for IoT datasets using various Machine Learning and Deep learning algorithms. These related work's inferences motivated the current research that utilizes an IoT-23 dataset for intrusion detection.

Sharafaldin et al. [15] created a reliable CICIDS2017 dataset that included seven common attack network flows in contrast to benign network traffic. They used seven common machine learning methods, namely K-Nearest Neighbor, Random Forest, ID3, Adaboost, Multilayer Perceptron, Naive-Bayes, and Quadratic Discriminant Analysis, to evaluate the performance and accuracy of the selected features. In addition, the new dataset is compared with publicly available datasets from 1998 to 2016 using 11 criteria that indicate common faults and complaints of the datasets that came before it. The findings of the comparison reveal that the recently developed dataset addresses all of the faults and issues. In order to evaluate the UNSW-NB15 and KDD99 datasets, Nour and Jill [16] presented a statistical analysis in which the UNSW-NB15 dataset would be split into a test set and a training set and evaluated from three perspectives: statistical analysis, feature correlation, and the complexity evaluation. Based on the findings, UNSW-NB15 may be relied upon as a valid dataset for testing both established and experimental IDS techniques. A unique intrusion detection system was suggested by Vijayanand et al. [17] using a genetic method to select features and several support vector machine classifiers. The suggested method, tested on the CICIDS2017 dataset, is predicated on picking the informative aspects for each category of assault rather than the common features of every attack. The experiments prove the unique approach works, and it successfully detects intrusions with a high rate of accuracy.

The genetic feature selection technique and multiple support vector machine classifiers were proposed by Vijayanand et al. [18] as a novel intrusion detection system. The proposed strategy relies on selecting the informative characteristics for each category of assault rather than the common traits of each attack, and it was evaluated using the CICIDS2017 dataset. The results show that the novel method is effective in detecting intrusions with a high degree of precision. The researchers Wang et al. [19] devised an approach for classifying malicious software traffic that was based on convolutional neural networks (CNN). The network traffic image is produced by mapping the traffic characteristics to pixels, and this image is then fed into the CNN so that traffic categorization may be accomplished using the data from the CNN. Kwon et al. [20] offers a Fully Convolutional Network (FCN) model as a result of their investigation into the deep learning model, which centers on data simplification, dimension reduction, classification, and other technologies. When compared to more conventional machine learning methods, the FCN model's value in analyzing network traffic is demonstrated.

Several ML algorithms, including the J48 Decision Tree, Random Forest, Multilayer Perception, Naive Bayes, and Bayes Network classifiers, were used by Obeidat et al. [21] on the KDD dataset. The attacks on the KDD dataset were detected and classified with the maximum accuracy using Random Forest in this study. There is evidence that DL algorithms have been used on this dataset as well. High precision has been achieved with the use of a deep neural network. Support vector machines (SVMs) and 1-nearest neighbor on features taken from the trained deep convolutional neural network (CNN) were used to construct few-shot intrusion detection by Chowdhury et al. [22]. However, there is network bias in the KDD99 dataset. About 78% and 75% of the records are duplicated in the train and test set, respectively, according to the analysis of the KDD99 dataset. Using Autoencoder, Zhang et al. [23] applied a DL model to the NSL-KDD dataset. The problem is that these data sets are nearly 20 years old. Aside from the problems inherent to the products themselves, they may lack assault scenarios that account for the development of new dangers.

Data for IP flow-based intrusion detection was reported by Sperotto et al. [24]. By identifying scans, worms, Botnets, and DoS attacks, it categorizes attacks and defenses. Normal user behavior, however, is not included. Normal user behavior was also accounted for in the network traffic included by Shiravi et al. [25]. But it couldn't talk to the remote server. This is crucial since most assaults in the actual world are launched from remote servers. As a result, an IDS may only be able to duplicate the attack scenario through its own internal server.

3   Dataset Description

New data for IoT network traffic machine learning techniques is the IoT-23 dataset. 23 IoT device malware captures and 3 benign traffic grabs total 760 million packets and 325 million annotated flows from 500 h of traffic. Avast Software in Prague funded the Stratosphere Laboratory, AIC group, FEL, and CTU University in the Czech Republic to develop the dataset in 2018–2019.

The IoT-23 dataset includes two new columns for network behavior description labels in both benign and malicious traffic flows. The labeling process involves the following steps:

1. An analyst manually analyzes the original.pcap file.
2. Analysis dashboard labels suspicious flows.
3. The analyst creates labels.csv with netflow labeling rules.
4. A Python script compares netflow data (conn.log) to labels.csv rules. Labels are added if netflow passes labeling criteria.
5. The conn.log.labeled file contains the original netflows and the updated labels from the human analyst study.

The Aposemat Project, Avast-AIC Laboratory in Prague, captured traffic from genuine IoT devices in 2018 and 2019 to build the IoT-23 dataset. Raspberry Pi malware developed the malicious scenarios. Mirai, Torii, Hide and Seek, and Hajime malware traffic is included.

4   System Framework

The proposed system framework consists of sections with feature selection and without feature selection phase in which an Extreme Random Tree (ERT) classifier-based feature importance is used to review the most relevant features present in the dataset (Fig. 1).

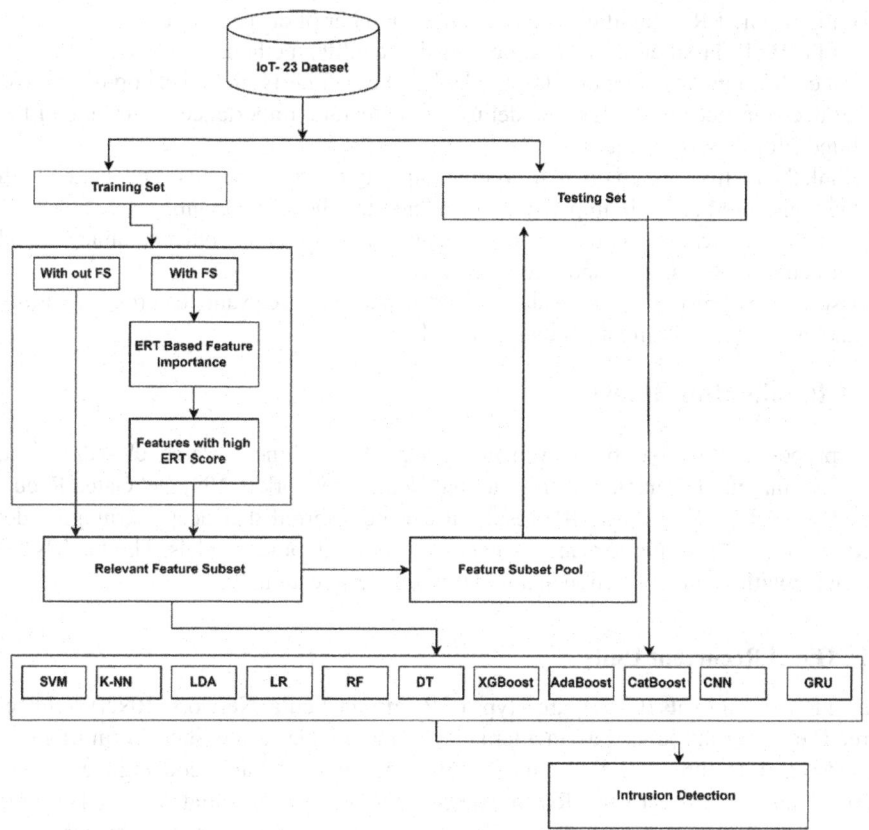

Fig. 1. System framework.

5   ERT Classifier Based Feature Importance

ERT (Extra Trees Classifier) is a tree-based machine learning algorithm that can be used for both classification and regression tasks. It creates multiple decision trees using random subsets of the data points and combines their predictions to make a final prediction. Feature importance can be calculated by looking at the reduction in impurity when a feature is used in a split. The more the impurity is reduced, the more important the feature is considered to be. Feature importance values can be obtained using the.feature_importances_ attribute of the fitted ERT classifier object. These values can be visualized using a bar plot or other suitable method for better understanding and interpretability. It's important to note that feature importance calculated by ERT is based on the reduction of impurity, it does not take into account the correlation between the features and the target variable [26].

The algorithm for ERT based feature importance can be summarized as follows:

- Split the input dataset into training and testing sets.
- Initialize the ERT classifier with a specified number of decision trees.
- Fit the ERT classifier to the training set using the.fit() method.
- For each feature in the input dataset, calculate the average reduction in impurity across all decision trees in the ERT model using the.feature_importances_ attribute of the fitted ERT classifier object.
- Rank the features based on their average impurity reduction values. The feature with the highest reduction in impurity is considered the most important.
- Visualize the feature importance values using a bar plot or any other suitable method for better understanding and interpretability.
- Use the feature importance values to select the most relevant features for further analysis or for building a prediction model.

6   Classification Phase

The proposed work has been evaluated using 10 Machine learning classifiers and deep learning models namely Convolutional Neural Network (CNN) and Gated Recurrent Unit (GRU) [27]. From the results it can be referred that deep learning model GRU achieved better performance when compared with other models. The models are evaluated with feature selection and without feature selection.

6.1   Gated Recurrent Units

Gated Recurrent Units (GRUs) are a type of Recurrent Neural Network (RNN) architecture. They were introduced as an alternative to the popular Long Short-Term Memory (LSTM) architecture, with the goal of simplifying the model and reducing the number of parameters. Like LSTMs, GRUs are designed to handle sequential data, such as time series, natural language, and speech. They use a gating mechanism to control the flow of information in the network. The gating mechanism consists of two gates: the update gate and the reset gate. The update gate controls how much of the previous state should be passed to the current state, while the reset gate controls how much of the previous state should be forgotten.

Gated Recurrent Units (GRUs) are a type of recurrent neural network (RNN) architecture that are used to process sequential data [28]. The main idea behind GRUs is to use gating mechanisms to control the flow of information through the network, allowing the model to better handle long-term dependencies in the data. A GRU unit can be represented mathematically as follows:

$$\text{Update gate: } z_t = sigmoid(W_z x_t + U_z h_{\{t-1\}} + b_z) \tag{1}$$

$$\text{Reset gate: } r_t = sigmoid(W_r x_t + U_r h_{\{t-1\}} + b_r \tag{2}$$

$$\text{Candidate activation: } h't = tanh(W_h x_t + r_t * U_h h_{\{t-1\}} + b_h) \tag{3}$$

$$\text{Hidden state: } h_t = (1 - z_t) * h_{\{t-1\}} + z_t * h'_t \tag{4}$$

where x_t is the input at time step t, h_t is the hidden state at time step t, $h_{\{t-1\}}$ is the hidden state at the previous time step, and W, U, b are the weight matrices and biases of the network. The sigmoid and tanh functions are used to control the update and reset gates, respectively.

The proposed work has been experimented with various optimizers for the GRU and the work has been evaluated both with feature selection and without feature selection.

The gates present in the GRU are typically trained using optimization algorithms such as stochastic gradient descent (SGD) or the Adam optimizer. In this proposed work the evaluation is performed with feature selection and without feature selection for the various optimizers for training the GRU network. The Table 1 outlines the list of optimizers used for training the GRU network.

Table 1. Type of Optimizers Used.

Optimizers Used	Definition
Ftrl	Follow-the-Regularized-Leader
Adadelta	Adaptive Delta
Adagrad	Adaptive Gradient Algorithm
RmsProp	Root Mean Squared Propagation
SGD	Stochastic Gradient Descent
Nadam	Nesterov-Accelerated Adaptive Moment Estimation
Adam	Adaptive Moment Estimation

7   Results and Discussions

The proposed work has been compared with 10 machine learning classifier and 2 deep learning models. The machine learning classifiers used in this work are namely, Support Vector Machine (SVM), K-Nearest Neighbor (K-NN), Linear Discriminant Analysis (LDA), Decision Tree (DT), Random Forest (RF), Naïve Bayes (NB), Logistic Regression (LR), XGBoost, AdaBoost, CatBoost, Convolutional Neural Network (CNN), Gated Recurrent Units (GRU). The accuracy comparison graph is as shown in Fig. 2.

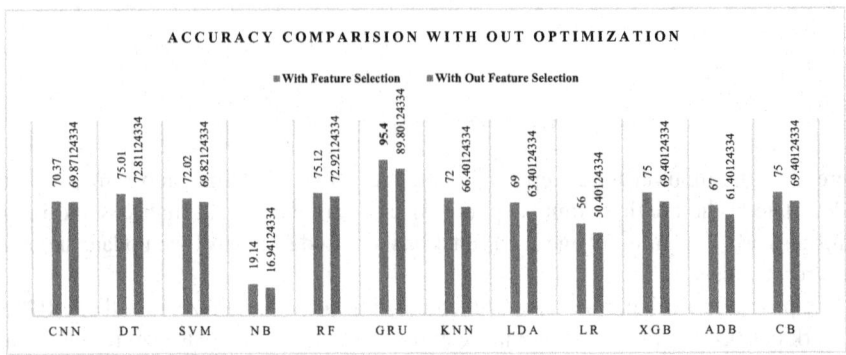

Fig. 2. Accuracy comparison without optimization.

The proposed work performs well with GRU and various optimizers are tested with feature selection and without feature selection. The Fig. 3 shows the performance of various optimizers with out feature selection along with the time. From the figure it can be inferred that the GRU with ADAM optimizer is performing well with an accuracy of 94.78% with a computation time of 1587.15 s.

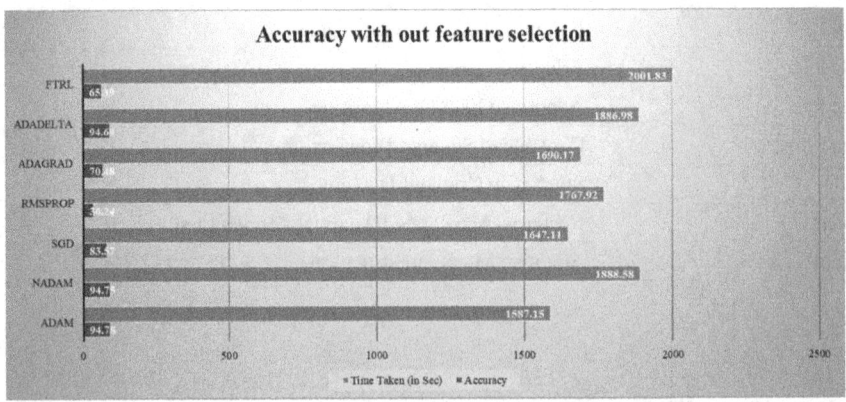

Fig. 3. Accuracy comparison without feature selection.

The Fig. 4 shows the performance of various optimizers with feature selection along with the time. From the figure it can be inferred that the GRU with ADAM optimizer is performing well with an accuracy of 96.89% with a computation time of 1011.15 s.

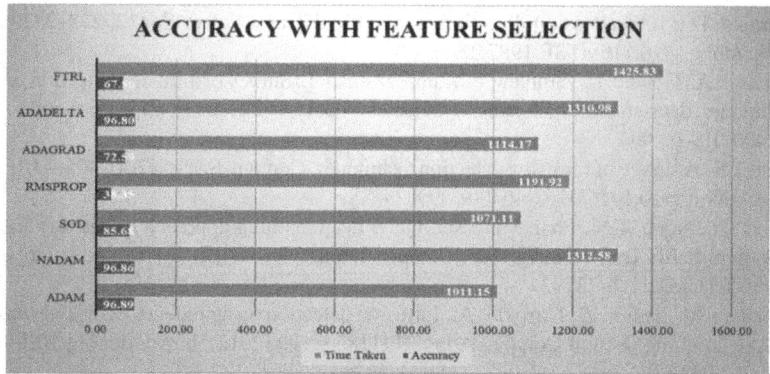

Fig. 4. Accuracy comparison with feature selection.

8 Conclusion and Future Scope

In conclusion, deep learning-based intrusion detection can be an effective approach for enhancing IoT security. The ability of deep learning models to process large amounts of data and identify patterns can be used to detect malicious activities in IoT networks. However, more research is needed to develop robust and efficient deep learning-based intrusion detection systems that can be deployed in resource-constrained IoT environments. Additionally, the development of techniques to improve the interpretability of deep learning models will be important for understanding the decision-making process of these models and for identifying potential vulnerabilities. Furthermore, the use of adversarial machine learning techniques to test the robustness of deep learning-based intrusion detection systems is an important area for future research. Finally, the integration of deep learning-based intrusion detection with other security measures, such as encryption and secure communication protocols, will be necessary to provide a comprehensive security solution for IoT networks.

References

1. Rosero, D.G., Sanabria, E., Díaz, N.L., Trujillo, C.L., Luna, A., Andrade, F.: Full-deployed energy management system tested in a microgrid cluster. Appl. Energy **334**, 120674 (2023). https://doi.org/10.1016/j.apenergy.2023.120674
2. Xia, F., Yang, L.T., Wang, L., Vinel, A.: Internet of things. Int. J. Commun. Syst. **25**(9), 1101 (2012). https://doi.org/10.1002/dac.2417
3. Villamil, S., Hernández, C., Tarazona, G.: An overview of internet of things. Telkomnika (Telecommun. Comput. Electron. Control) **18**(5), 2320–2327 (2020). https://doi.org/10.12928/telkomnika.v18i5.15911
4. Granell, C., Kamilaris, A., Kotsev, A., Ostermann, F.O., Trilles, S.: Internet of things. In: Manual of Digital Earth, pp. 387–423 (2020). https://doi.org/10.1007/978-981-32-9915-3
5. Gupta, R.K., Chawla, V., Pateriya, R.K., Shukla, P.K., Mahfoudh, S., Hussain Shah, S.B.: Improving collaborative intrusion detection system using blockchain and pluggable authentication modules for sustainable smart city. Sustainability **15**(3), 2133 (2023). https://doi.org/10.3390/su15032133

6. Denning, D.E.: An intrusion-detection model. IEEE Trans. Softw. Eng. **2**, 222–232 (1987). https://doi.org/10.1109/TSE.1987.232894
7. Khraisat, A., Gondal, I., Vamplew, P., Kamruzzaman, J.: Survey of intrusion detection systems: techniques, datasets and challenges. Cybersecurity **2**(1), 1–22 (2019). https://doi.org/10.1186/s42400-019-0038-7
8. Lunt, T.F.: A survey of intrusion detection techniques. Comput. Secur. **12**(4), 405–418 (1993). https://doi.org/10.1016/0167-4048(93)90029-5
9. Shone, N., Ngoc, T.N., Phai, V.D., Shi, Q.: A deep learning approach to network intrusion detection. IEEE Trans. Emerg. Top. Comput. Intell. **2**(1), 41–50 (2018). https://doi.org/10.1109/TETCI.2017.2772792
10. Sheikhan, M., Jadidi, Z., Farrokhi, A.: Intrusion detection using reduced-size RNN based on feature grouping. Neural Comput. Appl. **21**, 1185–1190 (2012). https://doi.org/10.1007/s00521-010-0487-0
11. Vinayakumar, R., Soman, K.P., Poornachandran, P.: Applying convolutional neural network for network intrusion detection. In: 2017 International Conference on Advances in Computing, Communications and Informatics (ICACCI), pp. 1222–1228. IEEE (2017). https://doi.org/10.1109/ICACCI.2017.8126009
12. Huang, S., Lei, K.: IGAN-IDS: an imbalanced generative adversarial network towards intrusion detection system in ad-hoc networks. Ad Hoc Netw. **105**, 102177 (2020). https://doi.org/10.1016/j.adhoc.2020.102177
13. Spadaccino, P., Cuomo, F.: Intrusion detection systems for IoT: opportunities and challenges offered by edge computing. arXiv preprint arXiv:2012.01174 (2020). https://doi.org/10.48550/arXiv.2012.01174
14. Vimal, S., Suresh, A., Subbulakshmi, P., Pradeepa, S., Kaliappan, M.: Edge computing-based intrusion detection system for smart cities development using iot in urban areas. In: Kanagachidambaresan, G.R., Maheswar, R., Manikandan, V., Ramakrishnan, K. (eds.) Internet of Things in Smart Technologies for Sustainable Urban Development. EICC, pp. 219–237. Springer, Cham (2020). https://doi.org/10.1007/978-3-030-34328-6_14
15. Sharafaldin, I., Lashkari, A.H., Ghorbani, A.A.: Toward generating a new intrusion detection dataset and intrusion traffic characterization. ICISSp **1**, 108–116 (2018). https://doi.org/10.5220/0006639801080116
16. Moustafa, N., Slay, J.: The evaluation of network anomaly detection systems: statistical analysis of the UNSW-NB15 data set and the comparison with the KDD99 data set. Inf. Secur. J. Glob. Perspect. **25**(1–3), 18–31 (2016). https://doi.org/10.1080/19393555.2015.1125974
17. Vijayanand, R., Devaraj, D., Kannapiran, B.: Intrusion detection system for wireless mesh network using multiple support vector machine classifiers with genetic-algorithm-based feature selection. Comput. Secur. **77**, 304–314 (2018). https://doi.org/10.1016/j.cose.2018.04.010
18. Vijayanand, R., Devaraj, D.: A novel feature selection method using whale optimization algorithm and genetic operators for intrusion detection system in wireless mesh network. IEEE Access **8**, 56847–56854 (2020). https://doi.org/10.1109/ACCESS.2020.2978035
19. Wang, W., et al.: HAST-IDS: learning hierarchical spatial-temporal features using deep neural networks to improve intrusion detection. IEEE Access **6**, 1792–1806 (2017). https://doi.org/10.1109/ACCESS.2017.2780250
20. Kwon, D., Kim, H., Kim, J., Suh, S.C., Kim, I., Kim, K.J.: A survey of deep learning-based network anomaly detection. Clust. Comput. **22**(1), 949–961 (2017). https://doi.org/10.1007/s10586-017-1117-8
21. Obeidat, I., Hamadneh, N., Alkasassbeh, M., Almseidin, M., AlZubi, M.: Intensive preprocessing of KDD cup 99 for network intrusion classification using machine learning techniques, pp. 70–84 (2019). https://doi.org/10.48550/arXiv.1805.10458

22. Chowdhury, Md.M.U., Hammond, F., Konowicz, G., Xin, C., Wu, H., Li, J.: A few-shot deep learning approach for improved intrusion detection. In: 2017 IEEE 8th Annual Ubiquitous Computing, Electronics and Mobile Communication Conference (UEMCON), pp. 456–462. IEEE (2017). https://doi.org/10.1109/UEMCON.2017.8249084

23. Zhang, C., Ruan, F., Yin, L., Chen, X., Zhai, L., Liu, F.: A deep learning approach for network intrusion detection based on NSL-KDD dataset. In: 2019 IEEE 13th International Conference on Anti-counterfeiting, Security, and Identification (ASID), pp. 41–45. IEEE (2019). https://doi.org/10.1109/ICASID.2019.8925239

24. Sperotto, A., Schaffrath, G., Sadre, R., Morariu, C., Pras, A., Stiller, B.: An overview of IP flow-based intrusion detection. IEEE Commun. Surv. Tutor. 12(3), 343–356 (2010) (2010). https://doi.org/10.1109/SURV.2010.032210.00054

25. Shiravi, A., Shiravi, H., Tavallaee, M., Ghorbani, A.A.: Toward developing a systematic approach to generate benchmark datasets for intrusion detection. Comput. Secur. 31(3), 357–374 (2012). https://doi.org/10.1016/j.cose.2011.12.012

26. Song, J., Zhu, Z., Price, C.: Feature grouping for intrusion detection based on mutual information. J. Commun. 9(12), 987–993 (2014). https://doi.org/10.12720/jcm.9.12.987-993

27. Altaha, M., Lee, J.-M., Aslam, M., Hong, S.: An autoencoder-based network intrusion detection system for the SCADA system. J. Commun. 16(6), 210–216 (2021). https://doi.org/10.12720/jcm.16.6.210-216

28. Bach, N.G., Hoang, L.H., Hai, T.H.: Improvement of K-nearest Neighbors (KNN) algorithm for network intrusion detection using Shannon-entropy. J. Commun. 16(8), 347–354 (2021). https://doi.org/10.12720/jcm.16.8.347-354

A Security and Vulnerability Assessment on Android Gambling Applications

Eric Blancaflor[✉], Robert Leyton-Pete J. Pastrana, Mark Joseph C. Sheng,
John Ray D. Tamayo, and Jairo Antonio M. Umali

School of Information Technology, Mapua University, Manila, Philippines
ebblancaflor@mapua.edu.ph

Abstract. Due to the pandemic, along with its lockdowns, a lot of entertainment places have temporarily or permanently closed down, like casinos. Since they were not allowed to go to these places, people have settled on online gambling. Like any other application, gambling apps require users to log in and input their information, such as credit card numbers, bank accounts, etc., to add money to their online gambling accounts. Some people also try to download these applications unofficially, thinking they might be able to gain an advantage that puts them at risk. In this study, the researchers aim to address the issues of downloading gambling applications from unofficial sources.

Keywords: Cybersecurity · Quick Android Review Kit (QARK) · Mobile Security Framework (MobSF) · Vulnerability Assessment · Security Assessment · Gambling Application · APK

1 Introduction

Gambling applications have been popular since the pandemic occurred in 2019, it has been a way to relieve stress and recreation for individuals since gambling sites and events have been closed. However, there is still money involved in online gambling with the use of different applications. In a recent report from The Business Research Company, the US has witnessed an increase of first-time online poker players by 255% since the coronavirus lockdowns began, while the poker industry experienced a 43% growth since April of 2020 [1]. As the gambling industry grows, an increase in security issues will also transpire. According to a 2022 Eclypses report, breaches in online sports betting have always been a substantial issue in the gambling industry. The online gambling industry has become extremely lucrative in the last few years and a major target for cybercriminals. This is because these websites and mobile applications are a gateway to the customers' credit cards, bank accounts, and other sensitive information [2]. In order to understand the threat these hacks pose and how to avoid becoming another victim, it is important to understand some of the recent attacks and the security vulnerabilities that were exploited [2].

Downloading applications from unknown sources have multiple threats that put devices and their users at risk. As reported by Bertel King, pumping all software from

© The Author(s), under exclusive license to Springer Nature Switzerland AG 2023
F. Neri et al. (Eds.): CCCE 2023, CCIS 1823, pp. 106–115, 2023.
https://doi.org/10.1007/978-3-031-35299-7_9

a single, trusted source is a way to keep devices secure. Developers create apps and upload them to the Play Store. Google checks them for viruses, malware, and anything else the company would consider malicious. Then it allows that app and updates to pass through to users. Out of the box, devices can only get affected by bad apps if the code manages to bypass Google's safeguards [3]. Application security tools reduce the risk from third-party sources. In Nadav Noy's article, taking a proactive and engaged approach to AppSec helps mitigate the risks involved with both internal and third-party applications. Code should constantly undergo application security testing. Application Security Testing (AST) is the process of testing an application in various ways to find security issues and then fixing them as part of the secure application development process [4].

The gambling application contains different payment methods included to buy things available in the application that the user can use for the character's improvement and for the game itself. There are several frequent attacks to be on the lookout for, although fraudsters never cease to develop innovative new ways to exploit the applications such as attacks involving payments wherein the online casinos and gambling applications or platforms are like digital wallets since the user must make deposits and/or withdrawals before playing the application [5]. Additionally, this process of the gaming application has an impact on the security issues of the gambling application if this is exploited and has not secured the environment. By continuous blocking of unauthorized gambling servers, gambling and gaming application developers typically create some gambling apps that are connected to several web servers that link to several third-party services, like as payment services, which would be exploited by illicit gambling apps to make the process of creating apps easier [6]. It is another issue that the researchers want to address in this study as the number of gambling applications has increased over time.

Quick Android Review Kit (QARK) is used to analyze the data of the applications in this study. QARK educates developers and information security personnel about potential risks related to Android application security, providing clear descriptions of issues and links to authoritative reference sources [7]. Another tool that is used is Mobile Security Framework (MobSF). It is a framework for automated mobile-app penetration testing, malware analysis, and security assessment that can analyze static and dynamic applications for Android, iOS, and Windows binaries [8].

1.1 Objective of the Study

The following are the objectives of the study:

- To analyze the security issues of the two gambling applications from third-party or unknown sources.
- To determine threats through reports from the QARK framework.
- To produce security reports through MobSF.

1.2 Scope and Limitation

The assessments will be done using Quick Android Review Kit (QARK) and Mobile Security Framework (MobSF), automated assessment tools for android applications. In using QARK, the researchers will perform the assessments on Kali Linux. A virtual machine will be used to avoid possible risks in the researcher's device and perform the assessments safely in a virtual environment. Two online gambling applications will be tested by downloading the Android Package or Android Package Kit (APK) versions of the applications from unofficial websites like APKPure or ApkTornado. QARK provides a report of possible vulnerabilities or threats of an application after the scanning. The researchers will also be testing the same applications using MobSF. The Mobile Security Framework will also be executed in Kali Linux for security purposes. After evaluation, MobSF produces a detailed report that includes the title of the threat, its severity, and the description of the said threat. The evaluation will be limited to gambling applications and will be only downloaded from sources that are not official or are not the primary application store. This is to ensure that no disturbances or exploits will be made on the official application. The results of the assessments will be used for educational purposes only.

2   Review of Related Literature and Studies

2.1   Vulnerabilities on the Mobile Phone and Applications

The numerous and potent capabilities that modern mobile devices have made it necessary in daily life and made it the target of attackers looking to acquire access to vital information of an individual [9]. According to OWASP [10], vulnerability is a gap or weak point in the application, which can be a design or programming error that permits an attacker to harm the application's stakeholders. There are many available tools and equipment used to know and address the issues related to the different vulnerability issues of a website, system, or application through the help of security and vulnerability assessment. According to NIST [11], security assessment is a process of testing and/or assessing a system's managerial, operational, and technical security controls to see how well they are being implemented, performing, and achieving the anticipated results about the system's security requirements. Furthermore, vulnerability assessment is the process of finding and detecting systemic gaps and these gaps offer a way to exploit the system [12].

2.2   Quick Android Review Kit (QARK) as a Vulnerability Scanner

Vulnerability scanners are automated tools for organizations to determine whether their networks, systems, and apps have any security flaws that could make them vulnerable to attack [13]. One of the vulnerability scanners available is the Quick Android Review Kit (QARK). It is a statistical analysis program created by LinkedIn Research that allows users to check APK files for a variety of known possible security flaws [14]. Using the Quick Android Review Kit (QARK), detecting Android devices' common vulnerabilities is rapidly and accurately found and provides documentation of any application vulnerabilities present for reviewers [15]. This tool will be used for the assessment of the gambling applications that were downloaded from third-party or unknown sources.

Based on Fig. 1, the QARK is being used to analyze the APK from dev. Studio and connects to another framework to get the result of the assessment made by both

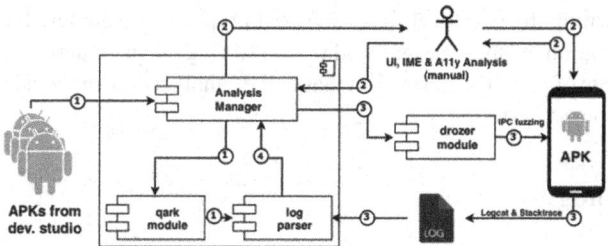

Fig. 1. Integration of QARK and other security frameworks for vulnerability assessment for APK.

frameworks used as an approach to analyzing vulnerabilities in the games from Google Play-published apps [14]. It shows that using the QARK can help to determine and analyze vulnerability to mobile applications since it is a great help to know those vulnerabilities in the APK to be addressed accordingly. Also, to improve the result of the analysis made by the other framework and to guide it to have better fast and more accurate results.

2.3 Mobile Security Framework (MobSF) as Another Tool for Security Scanning

Another security scanning used is Mobile Security Framework (MobSF) developed based on Python 3.7 which contains tools that do the static and dynamic analysis of an APK [16]. It was used to analyze the APK to know its behavior. This is one of the tools that will be adopted for testing the gambling application since it will determine the behavior of the APK whether it is a static or dynamic APK to gather information about its security issues.

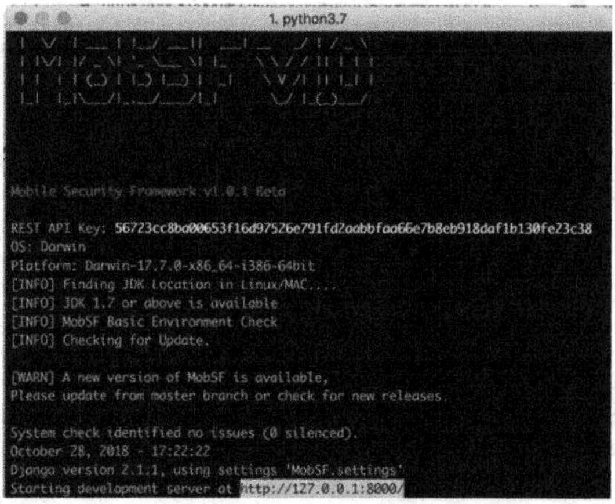

Fig. 2. Deployment of MobSF for statistics and dynamic analysis.

As shown in Fig. 2, the deployment of the MobSF to the mobile application is to inspect its security and to know the weak spot of the application to be able to address the issues after learning the result. MobSF analyzed properly and gathered all information about the application to have an organized way of storing the information in its local host. Also, the benefits of it to the information that will be analyzed to know the vulnerabilities which need to be addressed.

3   Methodology

The researchers conducted black box testing since they had no prior knowledge about penetration testing of the APK file and only added the inputs needed by the program and analyzed the outputs based on the generated reports. Additionally, the researchers used tools for the conducting of the testing which were the followings:

- Virtual Machine (VMware Workstation 16)
- Kali Linux 2022
- Quick Android Review Kit (QARK)
- Mobile Security Framework (MobSF)

The researchers analyzed different testing documents and videos available online to gather information about the process of QARK and MobSF that was done using Kali Linux. Additionally, vulnerability testing was done using QARK and MobSF for the additional testing which provides security testing to analyze the two (2) APK files downloaded from third-party or unknown sources to generate reports about the issues of the gambling applications. Both reports are used by the researchers to know the vulnerability and security issues with gambling applications.

For the vulnerability assessment using Quick Android Review Kit (QARK) testing;

- The researchers used a video [17] on how to conduct the QARK testing.
- Install the QARK by cloning it from GitHub to the virtual machine using the Kali Linux terminal.
- Install additional requirements (such as Python) to set up the virtual machine for the QARK.
- Open the QARK and add the APK file directory to analyze it.
- After analyzing the APK file, it generates a report consisting of different information about the issues in the APK.
- The report is used for the vulnerability assessment of the different APK files.

For the security assessment using Mobile Security Framework (MobSF) testing;

- The researchers analyzed different ways to conduct the MobSF testing which is available online and did it accordingly.
- Install the prerequisite of MobSF for static and dynamic analysis needed to set up the MobSF in the Kali Linux terminal.
- Cloning the MobSF from GitHub to the virtual machine and changing its directory to finish the last setup of the MobSF.
- Run the MobSF in the Kali Linux terminal to open its server.
- Open the browser and add the corresponding URL (http://127.0.0.1:8000).
- On the webpage of MobSF, the tester can add the corresponding APK file to analyze by uploading or adding its file directory.

- Both the terminal and the webpage of MobSF will work to analyze the APK until it is finished.
- The security reports will be available on the webpage or server of MobSF.
- The reports will be used for the security assessment of the APK files.

4 Results and Key Findings

4.1 APK 1 Using QARK

QARK scanner produced a report after scanning the application which allows the researchers to determine the possible vulnerabilities of the application. Based on the findings of the QARK scanner there are several threat results which will be discussed below.

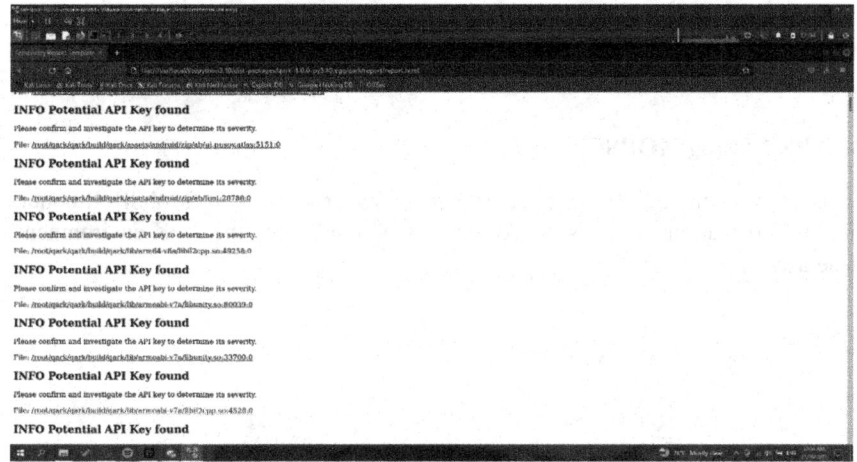

Fig. 3. INFO potential API key figure.

As shown in Fig. 3, there are potential information leaks of the data of the API calls of the application. This may enable access to projects that were not allowed or granted. This is a vulnerability of the application's permission. This can allow access to information inputted in the application where the privacy of users is violated. This threat disregards the permission to access the application and use resources that were denied by the users such as restricted data, system state, and user's personal information that they have registered in the application.

Presented in Fig. 4, logs may result in information being leaked from the application and detected in the APK file. Logs should only be present in the development phase of the application and not when the application is already developed or available.

4.2 APK 2 Using QARK

Same result as APK 1, There are potential information leaks of the data of the API calls of the application. This may enable access to projects that were not allowed or granted.

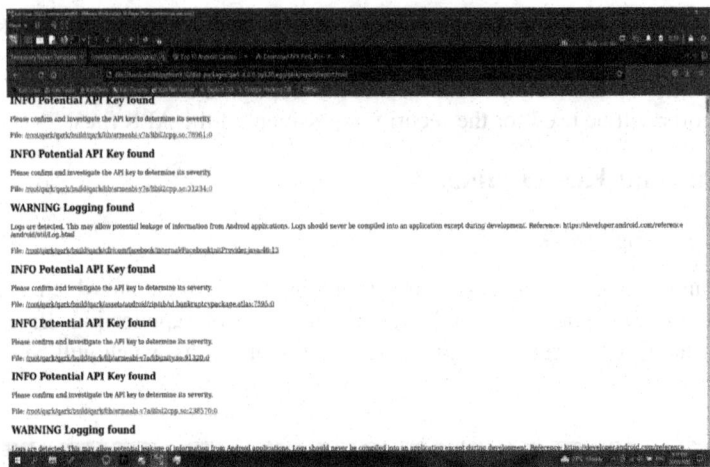

Fig. 4. WARNING logging found.

4.3 APK 1 Using MOBSF

The data was gathered after the security scanning process using MobSF. It contains different information to assess the APK to know how secure the APK environment is for the users.

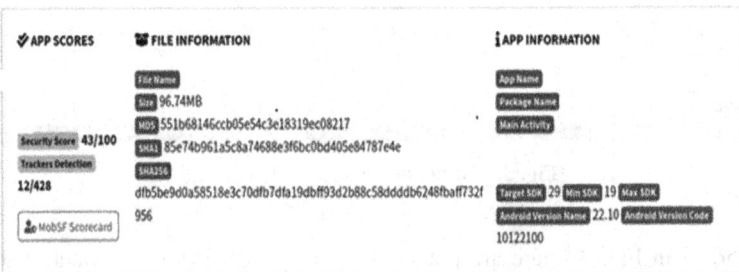

Fig. 5. APK 1 information and security score.

As shown in Fig. 5, MobSF provided insights about the basic information of the application, including APK 1 has a 43/100 security score and 12/428 tracker detection which means that the application has a lower security status based on the final scorecard provided by MobSF.

Based on Fig. 6, APK 1 has a high severity because its base configuration is not correctly configured to be secured to allow clear text traffic to all domains. Additionally, MobSF provides a description of the problem and why it has the conclusion of having a high severity on its network security.

For Fig. 7, MobSF shows that the application is potentially vulnerable to Janus Vulnerability which is prone to a signature scheme Android version 5.0 to 8.0 for v1 scheme

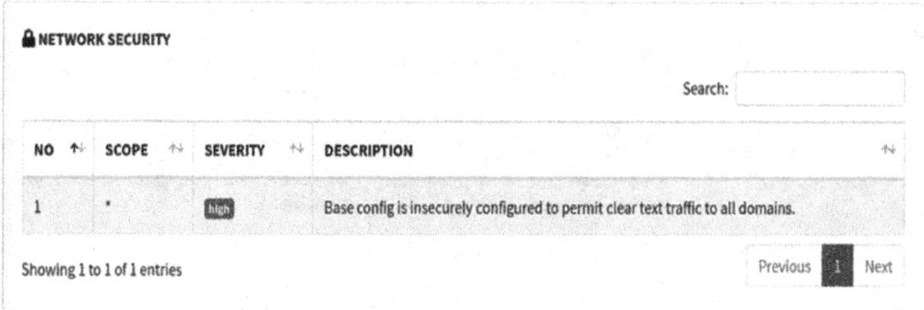

Fig. 6. APK 1 network security severity.

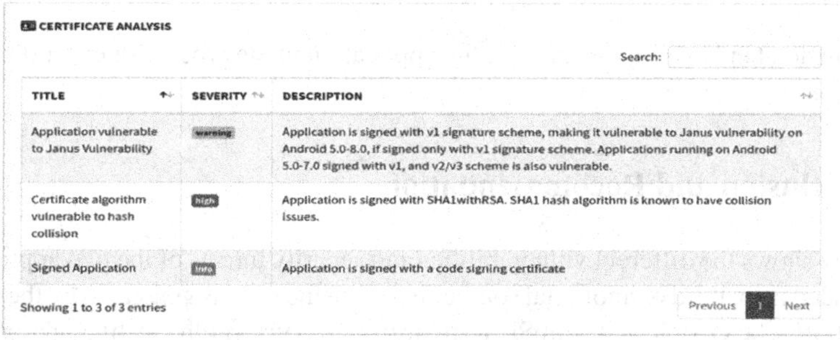

Fig. 7. APK 1 certificate analysis severity.

and Android version 5.0 to 7.0 for v1 and v2/v3 scheme that may exploit the attackers of the application. Additionally, the certificate algorithm is vulnerable because of the hash collisions and concluded as high severity that needs to be appropriately addressed to avoid attackers exploiting the sensitive user information provided in the application. The application has a signed application that requires code signing certificates that is understandable for the users of the application in how the process of execution of the application.

4.4 APK 2 Using MOBSF

Similar to the APK 1 process, APK 2 was processed using MobSF for the assessment. Additionally, it shows different basic information about the APK and the APK 2 security environment for the users who will be using the app.

MobSF provided insights about the basic information of the application including APK 2 has a 41/100 security score and 8/428 tracker detection which means that the application has a lower security status based on the final scorecard provided by MobSF.

APK 2 has a high severity because its base configuration is not correctly configured to be secured to allow clear text traffic to all domains. Additionally, MobSF describes the problem and why it has the conclusion of having a high severity on its network security.

Figure 8 shows that the application is potentially vulnerable to Janus Vulnerability which is prone to a signature scheme on the Android version 5.0 to 8.0 for v1 scheme and Android version 5.0 to 7.0 for v1 and v2/v3 scheme that may exploit the attackers of the application. Additionally, the signed application requires code signing certificates

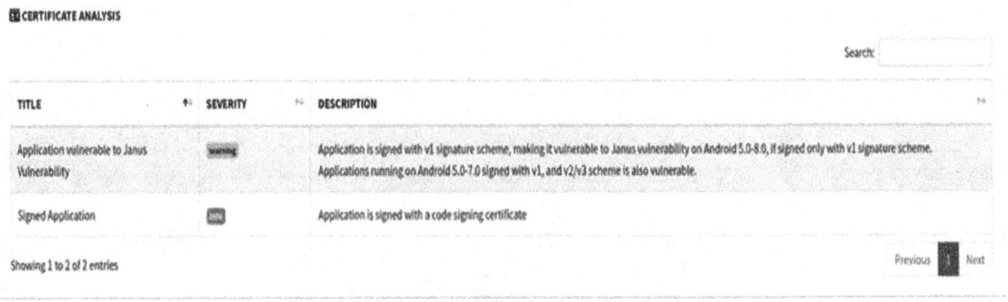

Fig. 8. APK 2 certificate analysis severity.

that are understandable for the users of the application in the process of execution of the application.

5 Conclusion and Recommendation

The study shows the different vulnerabilities and security threats of the two applications downloaded from third or unofficial sources based on the reports generated by the QARK and MobSF. The QARK and MobSF were utilized to perform the testing and based on the reports, many issues were found. Some were categorized as high severity, info, and warning severity based on vulnerability and security assessment reports. These findings are significant to emphasize the issues of gambling applications, especially those applications from unknown sources.

The study contributed to the understanding and knowledge of the respondents, and some factors are highly suggested for future awareness and use. The researchers recommend the followings:

- to be wary of downloading applications from unofficial sources since it might expose them to security threats.
- to check the reviews of the security status of gambling applications since it might pose a threat to their personal information.
- future researchers to utilize other security and vulnerability assessment tools in evaluating gambling applications downloaded from unofficial sources.

References

1. The Business Research Company. 2020. Online Gambling Platforms Gained Larger User Base With Pandemic As A Driver Of The Global Online Gaming Market 2020. Retrieved October 27, 2022, from https://www.globenewswire.com/news-release/2020/10/22/211 2777/0/en/Online-Gambling-Platforms-Gained-Larger-User-Base-With-Pandemic-As-A-Driver-Of-The-Global-Online-Gaming-Market-2020.html
2. Eclypses. 2022. Cyber Threats of Online Gambling | Five Online Gambling Platform Breaches to Learn From. Retrieved October 27, 2022, from https://eclypses.com/news/cyber-threats-of-online-gambling/

3. King, B.: 2016 Is It Safe to Install Android Apps from Unknown Sources? Retrieved October 27, 2022, from https://www.makeuseof.com/tag/safe-install-android-apps-unknown-sources/

4. Noy, N.: 2022. 5 Things You Need to Know About Application Security in DevOps Retrieved October 27, 2022. https://www.makeuseof.com/tag/safe-install-android-apps-unknown-sources/

5. Tanant, F.: 2022. Curious about online gambling fraud? all answers are here. SEON. Retrieved October 27, 2022. https://seon.io/resources/online-gambling-fraud/

6. Gao, Y., Wang, H., Li, L., Luo, X., Xu, G., Liu, X.: Demystifying illegal mobile gambling apps. In: Proceedings of the Web Conference 2021 (WWW 2021), pp. 1447–1458. Association for Computing Machinery, New York (2021). https://doi.org/10.1145/3442381.3449932

7. Srinivas. 2015. QARK – A tool for automated android app assessments. Retrieved October 27, 2022, from https://resources.infosecinstitute.com/topic/qark-a-tool-for-automated-android-app-assessments/

8. Joseph, R.B., Zibran, M.F., Eishita, F.Z.: Choosing the weapon: a comparative study of security analyzers for android applications. In: 2021 IEEE/ACIS 19th International Conference on Software Engineering Research, Management and Applications (SERA), pp. 51–57 (2019). https://doi.org/10.1109/SERA51205.2021.9509271

9. Argudo, A., López, G., Sánchez, F.: Privacy vulnerability analysis for android applications: a practical approach. In: 2017 Fourth International Conference on eDemocracy & eGovernment (ICEDEG), 2017, pp. 256–260 (2017). https://doi.org/10.1109/ICEDEG.2017.7962545

10. OWASP. Vulnerabilities. Retrieved October 6, 2022. https://owasp.org/www-community/vulnerabilities/

11. NIST. Security assessment. Retrieved October 6, 2022. https://csrc.nist.gov/glossary/term/security_assessment

12. Khera, Y., Kumar, D., Sujay, Garg, N.: Analysis and impact of vulnerability assessment and penetration testing. In: 2019 International Conference on Machine Learning, Big Data, Cloud and Parallel Computing (COMITCon), 2019, pp. 525–530, https://doi.org/10.1109/COMITCon.2019.8862224

13. Lucian Constantin. What are vulnerability scanners and how do they work? Retrieved October 27, 2022. https://www.csoonline.com/article/3537230/what-are-vulnerability-scanners-and-how-do-they-work.html

14. Kalysch, A., Schilling, J., Müller, T.: On the evolution of security issues in android app versions. In: Zhou, J., et al. (eds.) ACNS 2020. LNCS, vol. 12418. Springer, Cham (2020). DOI: https://doi.org/10.1007/978-3-030-61638-0_29

15. Albakri, A., et al.: Survey on Reverse-Engineering Tools for Android Mobile Devices, vol. 2022. Article ID 4908134 (2022). https://doi.org/10.1155/2022/4908134

16. Shahriar, H., Zhang, C., Talukder, M.A., Islam, S.: Mobile application security using static and dynamic analysis. In: Maleh, Y., Shojafar, M., Alazab, M., Baddi, Y. (eds.) Machine Intelligence and Big Data Analytics for Cybersecurity Applications. SCI, vol. 919, pp. 443–459. Springer, Cham (2021). https://doi.org/10.1007/978-3-030-57024-8_20

17. Briskinfosec. 2021. QARK - Quick Android Review Kit | TOD 174. Retrieved October 27, 2022. https://youtu.be/A_n-PmUjfts

A Compliance Based and Security Assessment of Bring Your Own Device (BYOD) in Organizations

Eric B. Blancaflor[(✉)] and Joel R. Hernandez

School of Information Technology, Mapua University, Metro Manila, Philippines
ebblancaflor@mapua.edu.ph

Abstract. Bring Your Own Device or simply called BYOD has quickly become the most major challenge for IT departments. The increasing number of employees using their own devices has caused the trend to become more prevalent. They are more likely to use their own laptops, tablets, and smart phones for work and personal use. With the number of tools that are available to communicate with different people, organizations data will be stored on various devices. Due to the increasing importance of security concerns, it is important that employees follow the policies and procedures of their organizations when it comes to bringing their own devices. A survey was conducted among organization employees who use the personal devices in the workplace. In addition, an interview was also conducted on the workplace to check how many personal devices are connected to the network. The results indicated that majority of the personal devices are connected to company network and that devices are using too much bandwidth. This study identifies the factors that influence an employee's decision to comply with the security policies of an organization and how the organizations implement the security policy and conduct a security assessment on the personal devices of the employees.

Keywords: BYOD · Security protocols · Wireless connection · Cybersecurity

1 Introduction

The BYOD strategy is considered a growth strategy for most organizations. It allows employees to bring their own devices into the organization to access various information resources to allow them to use their personal devices to access the organization's network and applications. Depending on the type of device they use, the BYOD strategy can be implemented in different ways. The rapid emergence and evolution of mobile technology is one of the factors that has contributed to the increasing number of organizations adopting this strategy [1]. However, the private network security level is typically lower than that of the public networks. Despite this, people still prefer to use their own devices due to various reasons. One of these is that it is easier to monitor their own agendas and reduce the amount of personal information that they have on their devices [2].

© The Author(s), under exclusive license to Springer Nature Switzerland AG 2023
F. Neri et al. (Eds.): CCCE 2023, CCIS 1823, pp. 116–127, 2023.
https://doi.org/10.1007/978-3-031-35299-7_10

One of the main advantages of BYOD is it allows employees to work faster and more efficiently as it allows them to respond quickly to calls and emails. However, it is important to keep in mind that it can expose the network to various threats, such as unauthorized access, theft, and the loss of data. Another important factor that the organizations should consider when it comes to implementing BYOD is ensuring that the devices are secure. One of the most common reasons why people lost their mobile devices is due to their portability. The combination of business and personal data on a mobile device poses a threat to organizations due to the possible unauthorized disclosure of sensitive information due to the files that are downloaded onto a device may not be protected from unauthorized access and are more vulnerable or infected with malware which may spread to the internal and external assets of an organization. In addition, BYOD devices may be located outside the organization, which means that organizations have less control over the devices and their users. Mobile devices may also be infected with malware, which could migrate to the company networks. Most mobile devices do not have antivirus software, which makes it easy for attackers to access and attack an organization's internal network and email system. Also, the traffic from web and email servers can be accessed remotely without being detected by firewalls [3].

Although BYOD policies provide numerous advantages, it also come with various privacy and security concerns. The nature of the devices that are used by employees makes BYOD an attractive target for security breaches. Employees who are working on their own devices, have higher risks of data leakage or unauthorized access. The pace of the BYOD environment is making it difficult to identify and manage the privacy and security challenges. Imminent effect as to security of mobile devices by mobile access has been observed, for example, mobile devices are often used by cybercriminals to develop and spread malicious code which may lead to attacks on these devices and may compromise the private information of users which the attackers can easily access even stored on a fixed-networked computers.

According to the 2020 AV Test, Windows is the most vulnerable operating system to malware attacks as compared to other operating systems tested which shows over almost 80% of all attacks and as showed in a data provided in 2019, over 114 million new pieces of malware were created, and 78.64% were directed toward Windows. Windows is also prone to security issues due to its numerous vulnerabilities. According to the database of known vulnerabilities known as the CVE database, Microsoft has over 660 security gaps, and 357 are attributed to Windows 10 which led it to be considered as the most insecure operating system in the world. This makes the cybercriminals take advantage easily of the system since almost half of the computers in the world run Windows 10. [4].

The popularity of Android has made it the most common operating system used on mobile devices. According to a threat study conducted by Kaspersky Lab in June 2017, many people are tempted to root their devices to gain unrestricted access to the files system, install different versions of the OS, and improve the performance of their device [5]. Android users should also be aware of the dangers associated with rooting their Android devices. This highlights the importance in considering the various security and privacy issues that affect both the organization and its employees.

1.1   Background of the Study

The rapid emergence and evolution of BYOD has created new security threats that can affect the way people work and personal lives. Mobile devices have become an integral part of today's work environment [6]. Due to the importance of protecting sensitive data, many organizations consider it a must to ensure that they are secure from the various threats that can affect their operations. While it is still challenging for them to secure their sensitive information on a regular basis, the emergence of BYOD has increased the threat that they face [7].

A security assessment is usually carried out to identify and implement the necessary security controls in organization. It can also be carried out to ensure that the design and implementation are secure. When employees attach their personal devices, such as laptops or mobile phones to organization's network or infrastructure, it makes sense to be concerned about their security. There is a risk that malware could easily transfer from the device to the organization's resources and computing devices. This type of data could also include company data that should only be kept private. Due to the nature of the data collected and stored on a device, it poses a threat to organizations due to the possible accidental disclosure of sensitive information. Therefore, it is important that the appropriate security measures are implemented when it comes to the sharing of business files on a device. In addition, the mobile device may contain malware that can infect the internal or business files of an organization. Bad things can happen if this kind of information is released to the public on a regular basis.

The goal of this study is to analyze the compliance of an organizations toward the implementation of policies and security measures that are designed to protect an organization's network and data from the threat of BYOD. In addition, this study aims to provide a policies and security measures that can be carried out through the implementation of security framework related to information security management.

1.2   Objective of the Study

The concept of BYOD allows employees to use their personal devices to access the organization's network and perform work-related tasks. It can include various types of devices such as tablets, smartphones, and USB drives. The main objective of the study is to assess the potential security risk of Bring Your Own Device (BYOD) in organizations. The following are the specific objectives, the goal is to: (a) to identify the risk of using personal devices in organization; (b) to explore if an organization has a security procedure or policy in the event of malicious activity or malware in personal devices; and (c) to identify vulnerabilities in a personal mobile device.

1.3   Scope of the Study

The study conducted a survey from different organizations. The researcher also interviewed a certain Senior Information Technology Officer of a company to get an understanding of their company's security measures and procedures, network infrastructure and preparedness. This study used different scanning tools such as Advanced IP scanner, Angry IP scanner, Zenmap and network monitoring tool to determine and identify the number of connected devices. In addition, the study conducted a full system scan on the personal devices in the workplace using Microsoft Safety Scanner in identifying possible threats.

2 Literature Review

2.1 BYOD – Compliance in Organization

According to the study of Cindy Zhiling Tu, Joni Adkins, and Gary Yu Zhao, due to the unique characteristics of BYOD, it is important that the management and employees understand the various security risks that can affect a company's operations. These risks can be minimized or eliminated by implementing the proper controls and strategies. Unfortunately, traditional security controls are not able to effectively address the challenges that BYOD presents. Beside adopting the proper technical measures, it is also important that the management and employees understand the various security measures that can be implemented to secure the use of BYOD. As part of their efforts to address the security concerns that can affect a company's operations, employees should additionally be taught how to comply with the policies [3].

A study conducted in 2015 by Trend Micro revealed that almost half of organizations that allow their employees to access their company's network using their personal devices experienced a data breach. The study also noted that device theft and loss were the most common reasons for these types of breaches. One of the most critical issues that organizations face when it comes to securing their networks is the exploitation of vulnerabilities in their BYOD devices [8].

Based on the study by Rmmd Pemarathna (2020), the BYOD concept allows employees to bring their own mobile devices into the workplace, which eliminates the need for them to carry around devices and provides them with better productivity. It can also lower the costs and improve the efficiency of the organization. Many organizations are also considering implementing BYOD programs, which allow employees to use their personal devices for work-related purposes. Most of the time, the companies allow BYOD due to its ability to allow employees to complete tasks outside the office. It is also allowed to participate in various communication and collaboration activities. The study also revealed that over 90% of organizations allow their employees to use their own mobile devices for work-related activities. Additionally, it was noted that this practice can have a significant impact on the efficiency of the organization and employee productivity [9].

2.2 BYOD Malware Attacks

Cybercriminals are reportedly increasing their efforts to infect users with malicious applications and text messages to steal their personal information, such as their passwords. According to experts at Proofpoint, the number of attempted mobile malware attacks doubled in the first few months of 2022. The company noted that the peak of these attacks occurred in February. Most mobile malware is focused on stealing login credentials and passwords for bank accounts and email accounts. However, it can also perform other invasive activities such as recording audio and video, monitoring your location, and wiping your data. As the evolution of mobile malware continues, more attacks are expected to utilize these advanced capabilities. The variant known as FluBot has been active since November 2020. It is designed to steal login details and passwords from various websites. It is also known as one of the most distinguished mobile malware variants. The ability to spread itself through various means, such as by sending SMS messages to its infected users' friends, is what sets FluBot apart from other variants. Proofpoint also identified the multiple mobile threats that can spread themselves through various means, such as by sending SMS messages to their infected users' friends. One of these is the multi-lingual malware known as the Moghau, which targets users around the world through fake landing pages that are designed to trick them into downloading a harmful program. Another is the TianySpy, which is a type of malware that targets Android and Apple users through messages sent from their mobile network operator [10].

According to the report of Atlas VPN, data breaches, ransomware, DDoS attacks, and viruses/malware were some of the most common cyber incidents experienced by government and business organizations from October 2019 to September 2020. Besides data breaches, other types of cyber threats are also common, such as viruses and malware infections in BYOD devices. According to a survey, over 43% of companies have experienced a virus or malware infection within their internal network in the past year. The concept of bringing your own device (BYOD) to work, which allows employees to use their devices for various tasks, poses additional risks due to the lack of proper security measures. Although it's beneficial to have a BYOD system, it remains questionable at best due to the lack of enterprise-level security measures. Around 60% of employees use personal devices for work, which was largely adopted following the COVID-19 pandemic [11].

According to the review of the researchers Mamoqenelo MOROLONG, Attlee GAMUNDANI, and Fungai BHUNU SHAVA (2019), personal computers are not always the target of attacks. Mobile devices are the most common target due to how they are used to performing various activities. Some of these include sending and receiving text messages and installing apps for financial transactions. The review states that attackers can use a mobile device to collect sensitive information, such as credit and debit card numbers, which they can use for monetary gain. With the emergence and evolution of multiple vulnerabilities in Android, researchers and practitioners have started to evaluate the security of the operating system [7].

3   Methodology

3.1   Research Design

The researcher both performed quantitative and qualitative research design. The survey design is used to examine audience compliance and opinions regarding the risk and threat of using their personal devices in workplace. A survey questionnaire was used in this study to collect relevant information for the study. Security Assessment was performed in a certain company in gathering relevant quantitative data. The respondents are workers from various sectors of organization coming from different departments. An actual interview to IT Senior Officer was also administered in their workplace and allow the researcher to conduct a wireless scan in identifying connected personal devices. In addition, a full system scan was conducted to one personal device in the workplace which examined the result whether the device will pass the security assessment and safe to connect in company network.

3.2 Security Assessment Process

To validate and test for any possible malware detection in BYOD, the researcher conducted a malware or virus scan and perform a network and security audit in a personal device of an employee to check for any open ports. A network scan was also conducted to determine how many personal devices are connected to the network. With security assessment, it will allow the researcher to detect a risk and vulnerabilities of a personal device (BYOD) in an organization.

4 Results and Key Findings

4.1 Survey Questionnaire Results

The result of survey questionnaire from different respondents has different insights with regard to the use and security concerns of the BYOD. The survey conducted is significant to understand and to know whether the respondents are willing to undergo company security assessment of their personal devices and to further see the compliance of the employee in implementing a stricter policy of Bring Your Own Device.

4.2 Respondents Profile

65.50% of the employees belongs to 25 to 39 age group while 25.90% from 18 to 24 and 8.60% from 40 to 60. Based on the results, it shows that majority of the respondents are from Private Companies with 74.10% followed by Government sector with 19%, Elementary and Senior High School with 5.2% and 1.7% from State Universities and College.

Most of the respondents' companies have more than 100 employees which is 60.30%. On the other hand, 25.95% of the respondents belong to a small size enterprise with 11 to 50 employees. 8.60% of the respondents are from the medium size enterprises which comprises of 51 to 100 employees and lastly, 5.2% of the respondents are from companies which only have 1 to 10 employees.

Respondents who have Intermediate Level Positions topped the survey results with 24.10%. Respondents who have positions as Rank-and-file employees follow the results with 20.70%. Respondents who belong to the Entry-level, Supervisory level, Managerial level, and Senior or Executive level then proceeded the results in a descending order. Only a few respondents are from the Government and Education Sectors, as well as elected officials.

As per the department where the respondents belong, the survey concluded that most came from the Accounting and Finance Department. Other departments only had less than 10% results.

4.3   BYOD at Work

According to the survey, more than majority of the respondents bring their own device at work. One of the questions also asked in the survey are the devices used while working, and mostly use both personal and work-issued devices, as opposed to using company issued devices or personal devices alone.

While most of the respondents which comprises of 91.50% have personal internet connection or mobile data on their personal devices, 66.10% or the majority connect their personal devices, which commonly are mobile phones and laptops, to their respective company's network. The reason for this is that it showed in the survey that majority of the respondents' companies have security policies in connecting personal devices. Only 1% of the respondents answered that their organization's network is unsecure. Accordingly, the 81.40% of the respondents said that they are still allowed to access their company's services, programs, and applications after office hours or even outside their workplace. These include their company email, shared files or folders, messaging software, company's own software or application, and company's database, in descending order according to the survey results.

The result showed the opinion of the respondents that even though they strongly agree in allowing to bring their personal devices at their workplaces, more than 30% of the respondents strongly disagree and another 30% are indifferent with regard to the installation of company's software in their personal devices. Small variances in the results were noticed that the respondents agree that their company should allow them to access the company resources through their personal devices. However, 40.7% of the respondents believe that their respective organizations should be liable in the event of data loss, theft or data corruption which may occur in their personal devices. On the other hand, the respondents have conflicted views as to their liability the company which were caused by their personal devices to the company resources in the event of being infected by virus or malware. This led the survey results to show that the organization should have a security assessment with the respondents' personal devices before connecting to their company network and that their company should implement a stricter policy regarding BYOD.

4.4   Security Assessment

The researcher conducted an interview with a Senior IT Officer of a certain company. The company is based in Asia, which provides delivery services through the web and mobile applications. As a strategic partner, the company has been able to help businesses solve their delivery problems. It has partnered with various organizations such as restaurants, retailers, and e-commerce companies, as well as with individuals. Through the Company's platform, it has been able to help businesses scale and provide customers with the best possible service.

The researcher conducted a wireless network scan to identify the numbers of available Access Point (AP) that are broadcasting in the company and to check for personal devices that are connected to the company network. Initial scan was conducted using the default Wireless Connection of Windows OS and insider, a network scanner for Wi-Fi. By using Wi-Fi scanner, the researcher found 12 SSID's that are currently broadcasting, 6 are official SSID of the company and the other are neighbor as shown in Fig. 1. The portion of the name on the scanned network was intentionally hidden and covered for confidentiality and privacy of the company.

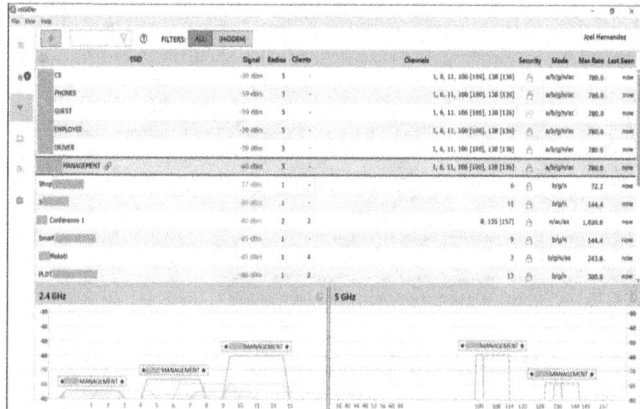

Fig. 1. inSSIDer Wi-Fi Scan Result.

The researcher requested with the Senior IT Officer for a possible report of devices in the company network to identify and filter what are the different devices specifically the Bring Your Device (BYOD) or personal devices that is connected to the network. As illustrated in Fig. 2. The captured data under the Device Inventory consist of Hardware Vendor, Software OS and the Device Status. The report generated came from company firewall.

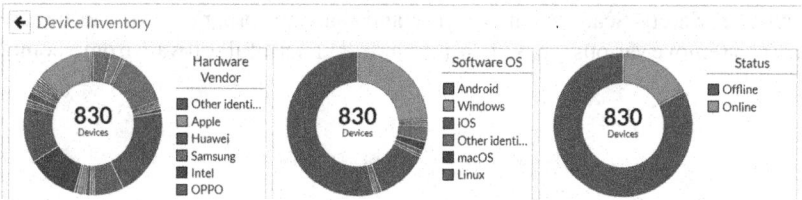

Fig. 2. Device inventory.

An inventory report provided includes the IP Address, Device (Hardware Vendor), SSID, Bandwidth Tx/Rx (device bandwidth consumption) and Signal Strength. Other details are hidden and covered for confidentiality and privacy of the company and to hide the identity of the devices. The purpose of this activity is to check and validate the result of Hardware inventory by scanning the LAN shows all connected devices through IP address.

The researcher conducted a network discovery and security assessment in a laptop (BYOD) of an employee to see if the device is compliant and will not harm the company network. The researcher uses a different tool to perform this task and the results of this scan can then be used to identify and resolve various security threats to other devices, the network and company resources. The device does not have Antivirus (AV) installed in the system and only Microsoft Defender is active and enabled in the system which can protect the device from various types of malwares, including viruses, ransomware, and Trojans.

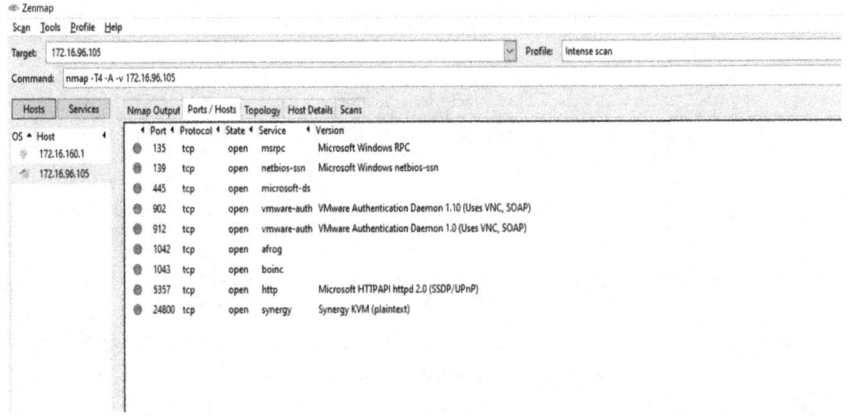

Fig. 3. Ports/Hosts.

Figure 4 shows the report after scan on a target device and the Ports /Hosts are displayed with State, Service and Version. As illustrated, it displays the open ports, the protocol and the type of services that is very vulnerable to any forms of attack and will be a great threat to the entire network as well as the other devices connected to the organizations network.

Microsoft Safety Scan which is a free anti-malware program that can be used to identify and remove various types of viruses and other harmful software from a computer.

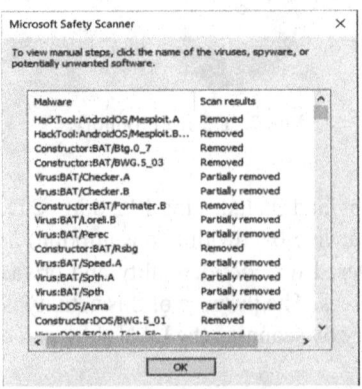

Fig. 4. Detailed results of the scan.

Figure 4 shows the detailed report and following malware of the infected files and the action taken by Microsoft Safety Scanner. This shows that the personal device (BYOD) is already infected without knowledge of the owner and also by the organization since all BYOD can connect to company network without being assessed or checked if the device will pass and comply with the organizations security policy.

The HackTool: AndroidOS/Mesploit. A is a threat that can arrive through different means, such as the installation of apps from unverified or unknown sources or the downloading of another malware. It can also exploit vulnerabilities in the device. After it has successfully launched and installed on the device, it can further compromise the device's network [12]. The Constructor: BAT/Formater. B is a program that is used to create a malware file automatically [13].

5   Discussion

BYOD provides employees with the control they need to manage their personal devices. The respondents agreed that the importance of the BYOD initiative is acknowledged by both the employees and organization. The result of security assessment conducted in one of the personal devices of the employee in an organization resulted to a non-compliant BYOD in a workplace. The target device posed a great security threat and risk in the organization due to the following factors: 1) the device has multiple open ports; 2) the device connects to the different unsecured networks such as coffee shop and other public places with WiFi; and 3) the device was already infected by different malware and viruses such as Trojan, Worms, etc. Insufficient security can also result in unauthorized access and modification of data due to negligent or careless actions. When it comes to personal devices, it is generally assumed that the users' actions will be transferred to business settings. There are also various compliance rules that companies must follow when it comes to using their personal devices.

Based on the respondents' feedback and security assessment of the personal device, the lack of policies and security controls, as well as the lack of awareness regarding the infected files and security threats of BYOD devices, were some of the factors that led to the issues related to compliance and vulnerabilities. Some of the security risks that can be considered are malwares and network attacks caused or brought by the personal devices. Employees might be exposing their own devices to security threats or vulnerabilities.

The development of an effective and efficient BYOD policy framework is a challenging task. It involves managing and controlling different types of devices using different hardware, manufacturer, operating systems and applications. One of the most important factors that an organization should consider when it comes to developing a comprehensive BYOD framework is the compliance and passing security assessment of the employee's personal devices.

6   Conclusion

The rise of the personal device has brought about the evolution of the distribution and use of technology. This is now known as the Bring Your Own Device or simply, BYOD, wherein employees can bring their own devices to work. It can help boost employee productivity and enable organizations to implement a more effective and efficient workforce.

Employees can gain more flexibility by being able to work more freely, and they also feel more comfortable using their own devices. While the concept of BYOD is beneficial for organizations, it should be considered carefully before implementing. However, there are some security concerns that prevent organizations from fully utilizing the BYOD model. The study explores the various risk and security challenges faced by organizations when it comes to implementing BYOD. One of the most critical components of this research is the detection of malicious activities and files are infected from personal devices. Despite the availability of the BYOD services, it is still risky to connect new personal devices to the company network.

An organization's information security policy should be the basis of a good BYOD policy. It should have a robust mechanism that can satisfy various requirements, such as secure network communications, identity access management (IAM), mobile device management (MDM), and enforcement of company policy. A balanced approach is also needed to ensure that the implementation of the policy is carried out properly. Having an information security policy should be the first step in establishing a good BYOD policy. In addition to regular security measures, there are also various measures that organizations can take to protect themselves from the risks associated with BYOD devices. The BYOD policy should clearly state how to separate the personal and business information. One step that an organization can take to protect its sensitive data is by implementing encryption. This method ensures that the data is protected from unauthorized access and use. Even in the worst-case scenarios such as theft, encryption can help protect the sensitive files.

Legislators, business, and individual consumers still require understanding the dangers they consent to each time they utilize a technology [14]. With this, mandatory security training should be required for all employees. This program will teach them how to identify and prevent security threats, as well as how to improve their personal safety. By following the policy, they can help prevent data breaches that can endanger the organization. The employees should immediately report the device to the company's IT department or manager if it's lost or stolen. The department should have the necessary tools and resources to perform the necessary steps, such as setting up a remote device lock and wiping the data. Employees should also be aware of the protocol for device loss or theft. If an employee is using public Wi-Fi or the Internet, they can be vulnerable to attackers who can access their business activities. To minimize the risk of being exploited, encourage them to connect their devices to a secure network, regardless of where they are located.

References

1. Alotaibi, B., Almagwashi, H.: A Review of BYOD Security Challenges, Solutions and Policy Best Practices. 23 August 2018 INSPEC Accession Number:18043305 (2018). https://doi.org/10.1109/CAIS.2018.8441967
2. Oktavia, T., Yanti, H.P., Meyliana: Security and Privacy Challenge in Bring Your Own Device Environment: A Systematic Literature Review. 18 May 2017 NSPEC Accession Number: 16885652 (2016) https://doi.org/10.1109/ICIMTech.2016.7930328
3. Tu, C.Z., Adkins, J., Zhao, G.Yu.: Complying with BYOD security policies: a moderation model based on protection motivation theory. J. Midwest Assoc. Inf. Syst. (JMWAIS), 2019(1), Article 2 (2019). https://aisel.aisnet.org/jmwais/vol2019/iss1/2

4. Cohen. 2020 Windows Computers Were Targets of 83% of All Malware Attacks in Q1 (2020). https://www.pcmag.com/news/windows-computers-account-for-83-of-all-malware-attacks-in-q1-2020

5. Casati, L., Visconti, A.: The Dangers of Rooting Data Leakage Detection in Android Applications. Volume 2018, Article ID 6020461, 9 p. (2017). https://doi.org/10.1155/2018/6020461

6. Rivadeneira, F.R., Rodriguez, G.D.: 2018 Bring Your Own Device (BYOD): a survey of threats and security management models. Int. J. Electron. Bus. **14**(2), 146 (2018). http://dx.doi.org/https://doi.org/10.1504/IJEB.2018.094862

7. Morolong, M., Gamundani, A.M., Shava, F.B.: Review of Sensitive Data Leakage through Android Applications in a Bring Your Own Device (BYOD) Workplace. Conference: 2019 IST-Africa Week Conference (IST-Africa). INSPEC Accession Number: 18849472 (2019). https://doi.org/10.23919/ISTAFRICA.2019.8764833

8. Kwietnia: Infosec Guide: Dealing with Threats to a Bring Your Own Device (BYOD) Environment (2017). https://www.trendmicro.com/vinfo/pl/security/news/cybercrime-and-digital-threats/-infosec-guide-bring-your-own-device-byod

9. Pemarathna, R.: Does Bring Your Own Device (BYOD) Adoption Impact on Work Performance? International Journal of Research and Innovation in Social Science (IJRISS) |Volume IV, Issue I, January 2020|ISSN 2454-6186 (2020). https://www.researchgate.net/publication/339302012_Does_Bring_Your_Own_Device_BYOD_Adoption_Impact_on_Work_Performance

10. Palmer, D.: Smartphone malware is on the rise, here's what to watch out for. (2022). https://www.zdnet.com/article/smartphone-malware-is-on-the-rise-heres-what-to-watch-out-for/

11. Wadhani, S.: Four Out of Five Businesses Were Cyber Threat Victims in the Last 12 Months (2020). https://www.spiceworks.com/it-security/vulnerability-management/news/four-out-of-five-businesses-were-cyber-threat-victims-in-the-last-12-months/

12. Microsoft. 2020. HackTool: AndroidOS/Mesploit. A. https://www.microsoft.com/en-us/wdsi/threats/malware-encyclopedia-description?name=HackTool%3aAndroidOS%2fMesploit.A&product=13

13. Microsoft. 2020. Threats. Available: https://www.microsoft.com/en-us/wdsi/threats/malware-encyclopediadescription?name=Constructor%3aBAT%2fFormater.B&product=13

14. Lange, T., Kettani, H.: On Security Challenges of Future Technologies. J. Commun. **14,** 1002–1008 (2019). https://doi.org/10.12720/jcm.14.11.1002-1008

Development of Home-Bot with Accident-Sensory System Using IOT and Android Application for Control

Patricia Louraine R. Caalim[1], Rexan Myrrh Calanao[1], Earl John Masaga[1], and Stephen Paul L. Alagao[2(✉)]

[1] Electronics Engineering Program, University of Mindanao, Davao City, Philippines
{p.caalim.476085,r.calanao.481367,
e.masaga.476136}@umindanao.edu.ph
[2] Computer Engineering Program, University of Mindanao, Davao City, Philippines
salagao@umindanao.edu.ph

Abstract. The Internet-of-Things (IoT) and robotics, individually, has been widely used in promoting better home security and quality oflife to its users. This study aims to design and develop a home-bot's accident-sensory system to monitor flame, smoke, and earthquake detection through the use of IoT; and develop an android application to monitor the transmitted data from the device (hubs) and even control the home-bot using Bluetooth. The real-time readings from all the hubs will be sent to the cloud Backend-as-a-Service real-time database. This data will then be reflected in real-time inthe android application and when either of the hubs detected any warning from any of the parameters, it will trigger an alarm to the hub itself and the android application for its user to react immediately. The system uses a nested loop algorithm to make sure that a detection is true in a span of time, together with a voting algorithm to avoid false alarms. The data gathered from the functionality tests was statistically analyzed using the confusion matrix method and resulted a 95.5% sensor accuracy for flame detection, 90.0% sensor accuracy for both the smoke and earthquake detection, and a successful transmission of data to the database to the android application in all the tests. The results showed significance accuracy and the system performed with accordance to its objectives and purpose.

Keywords: Internet-of-Things (IoT) · Google Firebase · Android Application · NodeMCU · Backend-as-a-Service (BaaS)

1 Introduction

Robotics is an emerging technology in the field of mechatronics. The robotic systems have reduced human efforts to do any work [1]. Also, smart phones have gradually turned into an all-purpose portable device and provided people's need. Moreover, an open-source platform Android has been widely used today. The mobile operated vehicle is a concept where a human being can control a vehicle by an android app by remote or

© The Author(s), under exclusive license to Springer Nature Switzerland AG 2023
F. Neri et al. (Eds.): CCCE 2023, CCIS 1823, pp. 128–140, 2023.
https://doi.org/10.1007/978-3-031-35299-7_11

wireless operation, without physically being seated inside it [2]. Nowadays, many studies have taken place to improve security systems. One of the major problems with regards to security system is the fire outbreak that may happen anywhere including houses, schools, offices, factories, and many other places [3]. Fire detection has become a very big issue which may cause severe damage [4]. The Bureau of Fire Protection (BFP) has recorded 1,758 fire incidents nationwide from January 1 to February 28 of last year, with death tolls at 22 and a property loss of over 1 billion pesos [5]. The top 3 causes of fire in the country are faulty electrical connections, lighted cigarette butts, and open flames. These top three causes are recorded to happen in residential areas, industrial and commercial buildings [6, 7]. On the other hand, earthquake is a natural occurrence which is due to the plate shifting of the earth that result to collision with other plates of the earth and raises a fault between the plates [8]. An earthquake is an unpredictable natural phenomenon which happens suddenly which we cannot stop but we can be alerted [9]. As per report, Philippines had encountered 882 earthquakes since February 2021. In which, 11 quakes were strong, 188 quakes were medium, and 683 quakes were weak [10]. And with Philippines being on the Pacific Ring of Fire, the island is one of the most natural hazard-prone areas. Hence, with the use of automation and sensors, it reduces the risk level of disasters such as fire accidents and earthquake. Moreover, the focus of discussion in this study is an application that tracks the fire detection sensor, smoke detection sensor, and earthquake detection sensor of the home bot.

In the study made by Kiran and Santhanalakshmi, a raspberry pi based remote controlled car using smartphone accelerometer was developed. The system's purpose is for surveillance, and it can be controlled anywhere in the world via smartphone accelerometer. Along with this, Wi-Fi communication between RC car and smartphone was established. The test was carried on the RC car with four directions which is the left, right, forward, and backward [11]. In a study made by U. Fatmawati, Hidayat, and Lelono, a proposed method using proportional derivative algorithm on the mobile robot fire detection and tracing was made. This method is a combination between wall tracing and fire tracing. Applying PD algorithm with TPA81 sensor increases the effectiveness of the robot in finding the fire spot and can detect fire directly [12]. In the study by Alphonsa et. al., they built an earthquake early warning system by means of an IOT in Wireless Sensor Network. The sensors are placed in the surface of the Earth. Once an earthquake occurs, both compression P wave and transverse S wave radiates outward the epicenter of the earth. The P wave, which travels fastest, trips the sensors, placed in the landscape. It causes early alert signals to be transfer ahead, giving humans and automated electronic system a warning to take precautionary actions. So that the public are warned earlier before any damage begins with the arrival of the slower but stronger S waves. The signal from each sensor which senses the P wave, and which has Zigbee transmitter transfers the alert signal to the gateway. Hence, early alert message is received by the people in terms of location, time, and other parameters [13].

This paper applies the Internet of Things (IoT) into the accident-sensory system of the home robot to monitor the gathered data from the different sensors used. The control of the home robot can be accessed through an android application which is deployed on the raspberry pi [14]. With this, a home robot with early detection of accidents or natural phenomenon including fire and earthquake will be accomplished. The system will provide an information that will allow individuals to protect their lives. An android application for the control of the home robot using raspberry pi will be developed as well to take prior action to a disaster [15].

The objectives of this study are to design a home robot's accident-sensory system; specifically, to create an IoT device that can detect fire, earthquake, and smoke; control the robot using an android application; perform a functionality test in terms of: sensor functionality, transmission and data logging, and the total device functionality.

The significance of this study is to prevent casualties and further damage from any hazardous or harmful events at home that is cause by accidents due to negligence or natural calamity. The homeowner can monitor the house in terms of the values read by the sensors using the android application.

The target beneficiaries for this paper are homeowners, or any small places as long as the microcontroller gets to connect to the internet through Wi-Fi. Hence, the android application will precisely monitor the sensors and control the motor driver of home bot which will greatly ensure the safety and condition of the home, with real time monitoring.

2 Materials and Methods

2.1 Conceptual Framework

Figure 1 shows the conceptual framework of this study. The input is from the fire detection sensor, smoke detection sensor, earthquake detection sensor, temperature detection sensor, online earthquake web data, and an android application joystick. The microcontroller then processes this data and perform the specified tasks that is programmed—data gathering, processing, and transmission. Lastly, with the data being processed and transmitted, the output is achieved.

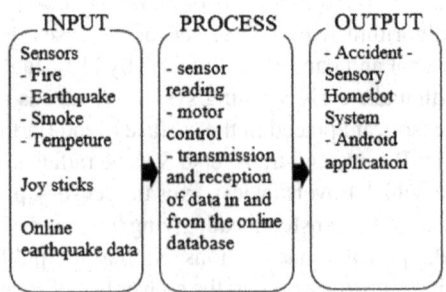

Fig. 1. Conceptual framework.

2.2 Functional Block Diagram

As shown in Fig. 2, the functional block diagram of this study's device follows a simple topology for data communication. The first procedure of the system involves two different kinds of sensor hubs, where one is equipped with a flame sensor, a smoke sensor, and a temperature sensor, and the second has only an additional earthquake sensor, and both hubs have a NodeMCU to serve as their microcontroller that will process and analyse the data from the 4 input sensors and then transmit the data into the cloud data storage which is the Google Firebase through the advantage of a built-in ESP8266 Wi-Fi module. The raspberry pi attached to the home bot will process and analyse the data together with the data from the online earthquake web scraping then transmit these data into the Google Firebase. Then, the android application will get the data from the database and presents these data together with the warning notification. The android app joystick will merely connect to the robot through Bluetooth, after which it will be processed and evaluated by the Arduino Mega to control the home robot's motor driver.

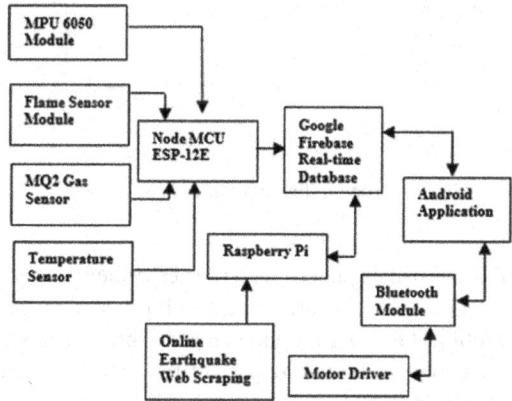

Fig. 2. Device functional block diagram.

2.3 Process Flow Diagram

There are a total of six hubs. All hubs have the same component content—flame sensor, smoke sensor, temperature sensor, node MCU, a buzzer, LED, reset and clear button, and a switch—except for the hub five and six where it has an earthquake sensor. It must be two hubs that contain an earthquake sensor since the validity and accuracy of an earthquake detection will rely on the detection of the two hubs using its voting algorithm together with the earthquake web scraped data.

Figure 3 shows the process flow diagram of all the hubs in terms of their temperature, flame and smoke monitoring. Each sensors have threshold value and the data that will be gathered from these sensors will be read and analyzed by the microcontroller. The microcontroller will then perform its specified task based on the data being processed from the sensors. As shown in the figure, these parameters follow the same nested loop algorithm which makes sure that a detection is true for two consecutive sensor reads in a span of two seconds.

As shown in Fig. 4, devices, or hubs 5 and 6 still follows the nested loop algorithm for the temperature, flame, and smoke sensors. But because of the additional earthquake detection sensors in both the hubs, they will follow a voting algorithm to further minimize

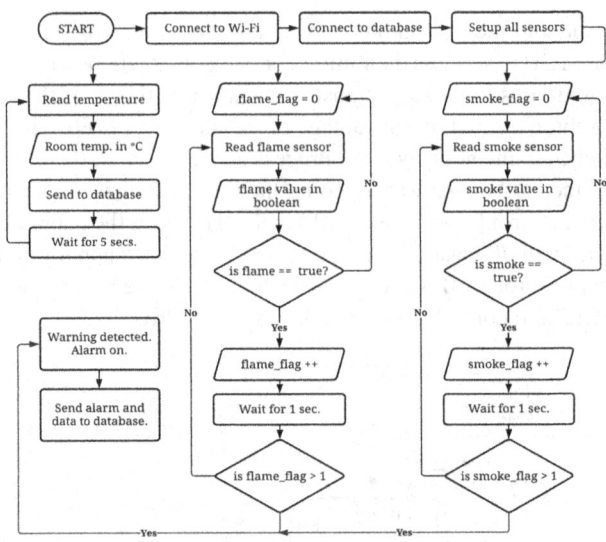

Fig. 3. Process flow diagram of the device.

the false detections that can still exist even with the performed proper sensor calibration. The hubs must be placed diagonally apart from each other to make sure they cover a different area for detection. Once hub 5 and 6 reads a ground movement, it performs a detection verification using the nested loop algorithm before sending an alarm to the database. Next, hub 5 then proceeds to loop on the database node where hub 6 sends its earthquake alarm for at least 5 s to make sure that hub 6 also detected the ground movement. If hub 6 never sent an alarm in that span of time, hub 5 concludes that it only detected a false alarm and goes back to monitoring all the sensor parameters. But when hub 6 also detected the ground movement, hub 5 then triggers all the alarms and waits for at least 30 min to receive all the earthquake details from the data that is web scraped from the PHIVOLCS website.

2.4 Software Integration

Nowadays, people start to use android phones in an ever-increasing manner especially with internet. In this study, android phones will be utilized to increase the efficiency of the performance of the device and an easy access for the homeowners that installs the application. The android application is connected to backend-as-a-service (BaaS) real-time database that will send real-time data updates and activates features especially the alarm system which will then be reflected and presented in the android application. The alarm features include phone notification, blinking of phone's flashlight, phone sound alarm, blinking of the LED on the activated hubs, and blinking of a button in the app's GUI. All of these are coded to catch the homeowner's attention in times of unexpected accident so an immediate action will be taken place.

The android application will be developed using Kodular, a free platform that allows you to create android apps with a blocks-type editor easily. First is the setting up the

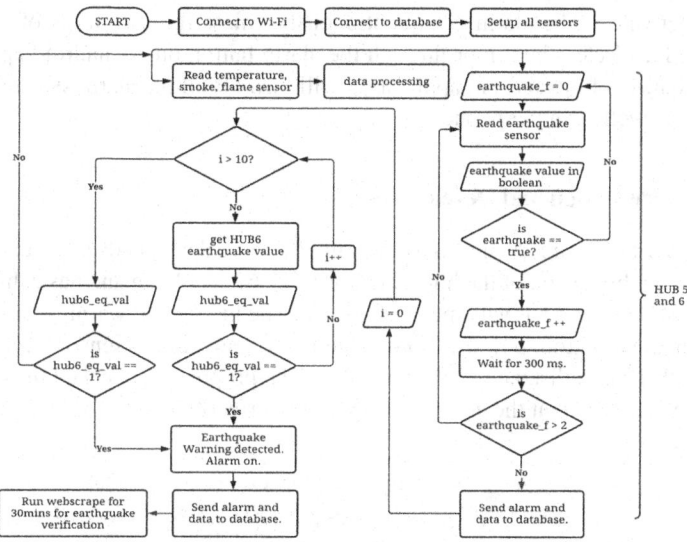

Fig. 4. Process flow diagram for earthquake detection.

android application. Install the provided APK file of the android application. After installing, open the app and log in using the provided credentials. This only allow users with the correct credentials to open the app. After successfully logging in, it will automatically redirect to the dashboard of the android app where the latest status of all the MCU hubs is displayed. Also in the dashboard, there is a button for the recent earthquake data web scraped from the official PHILVOCS website.

Next feature of the android app is the archive. It is divided into two buttons; the earthquake data and the sensor data. All the data gathered from the system are tabulated in this area. For the earthquake data archive, all the previous up until the recent data are tabulated with their time and date, location, magnitudes, and the link for the web scraped data from PHILVOCS. For the sensor data archive, all the data will also be tabulated with their time and date, name of the hubs, and the status of the hubs. The two features of the archive can be exported as an Excel file to be able to use Microsoft Excel's features and perform further data analysis.

The android application remote-control feature button is placed at the bottom part of the android application. The user needs to turn on the phone's Bluetooth then restart the android app. When the user clicks the remote-control button in the app, the app then attempts to connect to the MCU. When connected successfully, the joystick in the GUI of the app will be enabled and the user can now control the robot using the joystick in the app. The angle of the joystick will be sent to the MCU and will be analyzed by the MCU so the robot can go to direction that is controlled by the user.

The last feature of the android app is the alarm system. If the hub status is normal, it will just display on the dashboard saying the hubs are in good status. When a parameter is not within the threshold values, either one or all sensors, the android application activates the alarm system, which will continuously blink the phone's flashlight, the sound alarm of the phone, the red LED of each hub, and a warning notification indicating which hubs are being activated. The alarm goes continuously unless the user turns off the alarm through the hubs' clear button or through the alarm button on the android application. The alarm button of the android application will turn off all the alarm system in the all the devices as well as in the phone.

2.5 Hardware Design and Development

The 3D hub's enclosure design as shown in Fig. 5 is developed using the software SketchUp. The dimension of the hub's body is 94 × 69 × 60 mm and has a thickness of 2 mm. The MQ2 sensor has a diameter of 20 mm. The fire sensor, warning LED, buzzer, reset button, clear button, power LED indicator, and temperature sensor has a diameter of 6 mm with 10 mm distance from each other. The proponents checked the datasheets of the components so that the components will fit perfectly.

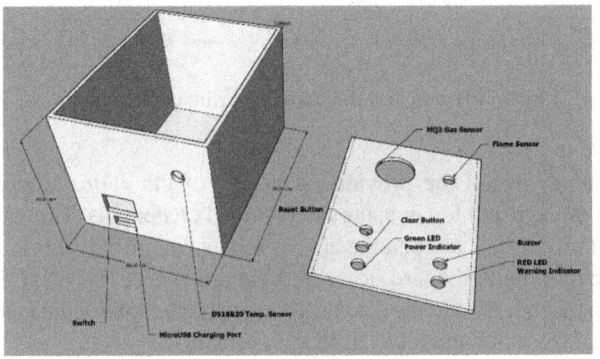

Fig. 5. Enclosure design of the hub.

For the components to be used, the proponents will use the NodeMCU ESP-12E open-source IoT platform. It includes firmware which runs on the ESP8266 Wi-Fi SoC from Expressive Systems which enables it to easily connect to a Wi-Fi connection. This microcontroller can establish a communication to the system's database which will handle the data transmission from all each hub. It is smaller in size, has better specifications than the UNO, and can be also coded in the Arduino IDE.

Flame detection will be done using the flame sensor based on the YG1006 sensor which is a high velocity and highly sensitive NPN silicon phototransistor. It is used to detect fire source that are in the wavelength in the range of 760 nm–1100 nm. On the other hand, MQ2 Gas sensor will be used to detect smoke as this sensor can detect or measure smoke and even gasses like LPG, Alcohol, Propane, Hydrogen, CO, and even methane concentrations anywhere from 200 to 10000 parts per million (ppm). For the earthquake detection, MPU6050 3-axis MEMS accelerometer will be used as it can measure angular momentum or rotation along all the three axes, the static acceleration due to gravity, as well as dynamic acceleration resulting from motion, shock, or vibration. The room temperature can be monitored using the DS18B20 which is a high-resolution temperature sensor up to an accuracy of ±0.5 °C from −10 °C to +85 °C.

The power source of each hub will be a single 18650 Lithium-ion battery which outputs 3.7 V and because the device runs on 5 V, a MT3608 DC-DC Step Up converter will be used to step up the voltage output from the battery to a constant 5 V. The battery is rechargeable and will be recharged using a TP4056 Micro USB charger module.

2.6 Conduct of Functionality Test

Functionality test will be conducted in a house for simulation and there will be at least six different sensor areas to where the hubs will be placed. To conduct its functionality, the surroundings will be manipulated so that the sensors will read the values of their respective parameters and trigger accordingly. The data that will be gathered are the detection of smoke, flame, and earthquake together with the date and time to when it was detected.

For the verification of the device's accuracy, data gathering and transmission capabilities, a series of tests will be made. All the fire detection sensor, smoke detection sensor, and the earthquake detection sensor will be tested for 20 samples each and tabulated accordingly to be statistically tested using the Confusion Matrix method. Also on the same method, the testing of the remote-control capability of the android application to control the home bot will be tested in the same procedure. Each test performed will also be checked if it successfully sent its data to the database. These data gathered will be statistically tested and verified for its accuracy and misclassification rate.

2.7 Confusion Matrix

To compute the system's accuracy, each sensor will be tested and have its accuracy solved using the Confusion Matrix method as shown in Table 1. Each sensor will undergo a test size of $N = 20$.

Table 1. Confusion matrix of the system parameters.

N	Total number of samples
True Positive (TP)	The cases in which the proponents predicted that there will be a detection and the system successfully read a detection
True Negative (TN)	The cases in which the proponents predicted that there will be no detection and the system did not read a detection
False Positive (FP)	The cases in which the proponents predicted that there will be a detection, but the system was not able to read a detection
False Negative (FP)	The cases in which the proponents predicted that there will be no detection, but the system read a detection

During the testing, the actual result will then be on the same row of the system's predicted result and its test number but on a different column labelled as the actual result. When the tests are finished, the sensor's accuracy (SA) will then be solved using Eq. 1.

$$SA = (TP + TN) / N \tag{1}$$

Misclassification rate (MR) or Error Rate, the rate of how often the system is wrong, can be solved using Eq. 2.

$$MR = (FP + FN) / N \qquad (2)$$

3   Results and Discussion

3.1   Device Fabrication and Functionality

Figure 6 shows the actual fabricated device of the study. It has a dimension of 94x69x60mm. All parts and components that were used in the device were made sure to fit in the printed enclosure and function in accordance with its purpose. The device can successfully connect to a wi-fi connection and to the BaaS real-time database of the whole system to perform its specified task—transmission and reception of data. In various conditions, the device have worked well, especially at times when the sensor thresholds were reached. The device responds real-time and its warning system—the buzzer and red LED turned on and a notification showed up in user's phone screen—was functioning well and stopped when the clear button in the device and/or android application's warning button was pushed. In addition, the device works well with a single 18650 Lithium-ion battery as its power source. This 3.7-V battery has its output boosted to 5 V using a MT3608 DC-DC Step Up converter. The battery inside is rechargeable and was charged using the TP4056 Micro USB charger. All the 6 hubs or devices, together with its android application, were working accordingly.

The android application, once installed, asks for the user credentials on the first log in. Once the log in is successful, it will redirect the user to the dashboard as shown in Fig. 7 where the user can monitor the real-time status of each hub.

Fig. 6. Actual device.

The alarm system of the android application—blinking of phone's flashlight, loud alarm sound, phone notification, dashboard update; worked perfectly well together with the device's buzzer and red LED alarm turning on when detected a simulated fire and/or smoke. On the other hand, as the new warning alert was received by the android application, it was also updated on the sensor archive page of the system. The latest data received

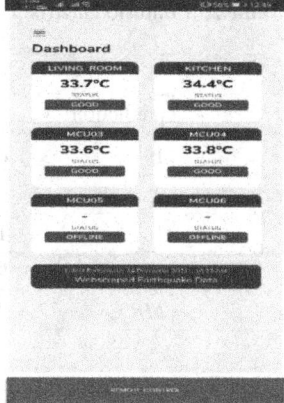

Fig. 7. Android application dashboard.

was added on the archive together with its date, time and what parameter was detected. The earthquake detection of the system followed the individual nested loop algorithm for the single hub detection and made sure a detection was true in the area of testing using the voting algorithm programmed to both hubs 5 and 6. During the earthquake simulation, the two hubs must both detect the ground or surface movement before alarming the user. The alarms were triggered, and the python script was also pinged to start web scraping on the PHIVOLCS website. But because this was just a simulated earthquake, no data was found in the PHIVOLCS website and after 30 min, the detection was considered a false alarm and notified the user for the updated. On the other hand, the researchers turn on the hubs for weeks and on one case, the hubs successfully detected an earthquake and after less than 30 min, the data from PHIVOLCS was delivered to the android application with all the details of the detection. After the users turned off the alarm system, the PHIVOLCS data reflected on the application and also on the earthquake archive page. The remote-control capability of the android application to control the movements of the home bot was also working well as intended. The Bluetooth connection enabled a mid-range control but real-time control of the home bot. The status of the connection has been established and the home bot can be controlled using the joystick on the right of the screen.

3.2 Data Gathering and Statistical Analysis

Computation of the system's accuracy was done by testing the multiple hubs in a set size of 20 using the confusion matrix method. This was done for all the three parameters—flame, smoke, and earthquake detection.

As shown in Table 2, during the simulation with an N of 20 for each sensor parameters, the flame detection has a sensor accuracy of 95.5% and misclassification rate of 0.05%, while the smoke and earthquake detection has 90.0% of sensor accuracy and 1% of misclassification rate. On the other hand, the remote control worked with an accuracy of 100% at all tests from establishing a Bluetooth connection up to testing the joystick movement to the home bot's movement.

Table 2. Confusion matrix.

Flame Detection			
20		prediction	
		1	0
Actual	**1**	9	0
	0	1	10
		SA	**95.50%**
		MR	**0.50%**

Smoke Detection			
20		prediction	
		1	**0**
Actual	**1**	8	0
	0	2	10
		SA	**90.00%**
		MR	**1.00%**

Earthquake Detection			
20		prediction	
		1	**0**
Actual	**1**	8	0
	0	2	10
		SA	**90.00%**
		MR	**1.00%**

Remote Control			
20		prediction	
		1	**0**
Actual	**1**	10	0
	0	0	10
		SA	**100.00%**
		MR	**0.00%**

For data transmission and reception of the system, across all the total of 60 tests performed from all the sensor parameters, all the alerts detected were successfully sent and received by the database and displayed on the android application. Also, all the data gathered reflected on the archive. For the reception of data, the earthquake data from the PHIVOLCS were always received and also tabulated on its respective archive page. Alarms were also successfully turned off on both the application and the hubs from either pressing the hub's clear button or the clear alert button in the android application.

4   Conclusion and Future Works

The designed and developed system in this research showed accuracy, functionality, and successful transmission, reception of data from all the hubs to the database up to the hands of the end user through the android application. Each sensor and feature of the system performed to its best condition providing a significant accuracy based on the statistical analysis made on the data gathered from the series of tests. The researchers conclude that this system has achieved its objectives to work as an accident-sensory system to monitor a home or establishment remotely in real-time based on the results presented from the system functionality tests.

The researchers, for future works, recommends to further improve the existing system through machine learning for better monitoring through a web camera. Also, an addition of an offline alert system to send alert through SMS will enhance the transmission and reception of data.

References

1. Singh, D., Agrawal, M., Patel, V., Das, E.: Arduino and sensors based fire fighting robot. J. Instrumentation Control Eng. **6**(2), April 2018
2. Shelki, D., Maniyar, A., Pawar, P.: Android app based car controller using bluetooth communication. Int. J. Adv. Res. Comput. Sci. Softw. Eng. **7**(5), May 2017. ISSN: 2277 128X
3. Khalaf, O., Abdulsahin, G., Zghair, N.: IOT fire detection system using sensor with Arduino. Defence Sci. J., August 2019
4. Saeed, F., Paul, A., Rheman, A., Hong, W., Seo, H.: IoT-based intelligent modeling of smart home environment for fire prevention and safety. J. Sensor Actuator Networks, February 2018
5. ABS-CBN News, "In the Know: The top 3 causes of fire in PH," ABS-CBN News, para. 3, Dec. 20, 2019. https://news.abs-cbn.com/news/03/01/18/in-the-know-the-top-3-causes-of-fire-in-ph. Accessed March 5, 2021
6. [Mayuga, J.L.: Tragedy of fires: Death and destruction in the Philippines. Business Mirror, March 21, 2018. https://businessmirror.com.ph/2018/03/21/tragedy-of-fires-death-and-destruction-in-the-philippines/ [Accessed March 5, 2021]
7. Kiran, V., Santhanalakshmi, S.: Raspberry Pi based Remote Controlled Car using Smartphone Accelerometer. The Institute of Electrical and Electronics Engineers, Inc., pp. 1536–1542, February 2020. https://doi.org/10.1109/ICCES45898.2019.9002079
8. Hammad-u-Salam, Z., Memon, S., Das, L., Hussain, A., Shah, R., Memon, N.: Sensor based survival detection system in earthquake disaster zones. IJCSNS Int. J. Comput. Sci. Network Secur. **18**(5), May 2018
9. Dutta, P.: Earthquake alarm detector microcontroller based circuit for issuing warning for vibration in steel foundations. IJMEC **7**(26), 3582–3594 (2017)
10. Volcano Discovery, "Latest earthquakes in or near the Philippines," Volcano Discovery, March 4, 2021. https://www.volcanodiscovery.com/earthquakes/philippines.html [Accessed March 5, 2021]
11. Gaur, A., et al.: Fire sensing technologies: a review. IEEE Sensors J., 1 (2019). doi:https://doi.org/10.1109/jsen.2019.2894665
12. Fatmawati, U., Hidayat, W., Lelono, D.: Novel method of mobile robot fire detection and tracing using proportional Derivative (PD) algorithm. Appl. Mech. Mater. Trans. Tech. Publications, Switzerland, **771**, 68–71 (2015)

13. Alphonsa, A., Ravi, G.: Earthquake early warning system by IOT using wireless sensor networks. Piscataway: The Institute of Electrical and Electronics Engineers, Inc. (IEEE). Retrieved from (2016). https://www.proquest.com/conference-papers-proceedings/earthquake-early-warning-system-iot-using/docview/1830961332/se-2?accountid=31259
14. Shirolkar, R., Dhongade, A., Datar, R., Behere, G.: Self-driving autonomous car using raspberry Pi. Int. J. Eng. Res. Technol. (IJERT). ISSN: 2278-0181 **8**(05), May-2019
15. Ibrahim, D.: Low-power early forest fire detection and warning system. Indian J. Sci. Technol. **13**(03), 286–298 (2020)

AI-Based System Model and Algorithm

Application of Machine Learning in Predicting Crime Links on Specialized Features

Omobayo A. Esan[1]([✉]) and Isaac O. Osunmakinde[2]

[1] School of Computing, College of Science, Engineering, and Technology,
University of South Africa, Pretoria, South Africa
[2] Computer Science Department, College of Science, Engineering, and Technology,
Norfolk State University, Norfolk, VA, USA
ioosunmakinde@nsu.edu

Abstract. Crime has negatively impacted the individual's life and the nation's economic growth. Currently, manual human assessments are used by security operatives to analyze the relationship between crime location and crime types from huge crime datasets, which are tedious and overwhelming. Hence, subject the criminal prediction results to errors. While many researchers make use of static crime dataset features for prediction which affects the prediction results, fewer approaches have focused on using crime dynamic features to address this lacuna. This research develops a machine learning-ensemble model based on dynamic crime features to address the issue of inaccuracy affecting crime prediction systems. Experiments were conducted on an Africa-based police crime data repository. Based on the experimental results, the proposed model outperforms the state of art models in terms of average precision, F1-score, and accuracy with 0.97, 0.97, and 97.03% respectively. The deployment of this proposed model in a complex environment can help security personnel to solve crime accurately and have a better response towards criminal activities.

Keywords: Classifiers · Crime · Link Prediction · Ensemble · Derived Features

1 Introduction

Criminal activities are present in every aspect of human life in the world, and it has negative consequences on people's lives as well as on the socioeconomic development of a nation [1]. The continuous increase in this act has become a major concern for almost all countries, and the effective mechanism to curb this issue is still a challenging issue.

The availability of a high volume of crime data is often used to understand the relationship between crime and crime types, however, analyzing the increasing volume of crime data manually can be tiring and overwhelming, and therefore, exposes the security system that needed the information for decision-making purposes to predictive errors and misinterpretation [2].

F. Neri et al. (Eds.): CCCE 2023, CCIS 1823, pp. 143–157, 2023.
https://doi.org/10.1007/978-3-031-35299-7_12

Traditionally, criminologist study and analyze historical crime data by focusing on sociological and psychological theories to link crime locations and the crime types. However, utilizing such strategies can be biased [3]. The existing crime prediction techniques have contributed to the use of crime dataset baseline features but they are restricted with accurate prediction performance, because they often rely on static crime data features for prediction [4]. Thus, imperative to develop an intelligent predictive security system that integrate dynamic crime features to predict the possible future link between the crime locations and crime types before the crime occurs. This study, therefore, develops a security machine learning model on derived criminal periodic features and an ensemble model that can intelligently predict the possibility of links link between crime locations and crime types, helping police to take necessary and early decisions in decreasing the crime rates.

The contributions of this paper are as follows:

- The integration of dynamic (periodic) features to the existing baseline crime dataset, to improve the performance of the link prediction systems.
- The development of the ensemble models which eradicates the worries of security operatives in decision-making processes.
- Extensive experimental evaluations of the proposed model on publicly available Africa-based crime data, and benchmark the model performances with related state-of-the-art prediction models.

The deployment of this proposed model in a complex environment is an emerging area. The remainder of this paper is arranged in the following order: Sect. 2 provides related works and the theoretical background of the proposed model. Section 3 presents the proposed methodology; Sect. 4 discusses various experiments and evaluations of the model. The concluding remarks are shared in Sect. 5.

2 Related Works

The research in [5] addressed the issue of difficulties in crime prediction that arises due to the subject's randomness by using classification techniques such as Naïve Bayes (NB), Random Forest (RF), and Gradient Boosting decision tree on the San Francisco crime dataset. The experimental results obtained show that the Gradient Boost Decision Tree (GBDT) gives 98.5%, NB achieved an accuracy of 65.82%, and RF gives 63.43%. The limitation of their approach is that their experiment did not show how the model improved crime prediction performance.

Also, work done in [6] addressed the challenging issue of classification of crime location and prediction utilizing deep learning architecture which involves Convolutional Neural Networks and Long Short-Term Memory (CNN-LSTM) for prediction. Although the authors indicated that their approach achieved satisfactory results, the technique is computationally complex and required domain experts' knowledge for its implementation in real-life scenarios.

Authors in [7] address the problem of resolving crimes faster and more effectively to understand criminal behavioral patterns by using a four-learning technique such as k-Nearest Neighbour (k-NN), Support Vector Machine (SVM), and XGBoost. Their experimental result shows that the proposed approach achieves up to 70% accuracy and a precision of 72%. The explanation regarding the feature extraction for the experiment was not given. Uncovering hidden patterns and analyzing the large data to prevent crime was introduced in [8]using a Multinomial Logistic Regression (MLR) method. The model was able to predict the weekdays, districts, and the likely hours the crime incident can occur. However, due to the limited crime attributes used during the implementation, improved accuracy could not be achieved. The research mainly adopts three different prediction models namely: the SVM, k-NN, and DT models to form models to forecast future links between crime types and crime locations. The following section explains the proposed detailed methodology for crime link prediction.

2.1 Selected Theoretical Techniques

The research mainly adopts three different prediction models namely: SVM, k-NN, and DT model to form models to forecast future links between crime types and crime locations. These models are briefly discussed in subsequent sections.

3 Proposed Crime Link Prediction by Classification

The system architecture is designed to improve the link prediction between crime locations and crime types in a crowded environment, and this is divided into three stages, including the data acquisition stage, data pre-processing stage, and prediction stage, see Fig. 1.

3.1 Stage 1: Data Acquisition

To ensure the universality of the classifiers, the real-life crime dataset is obtained and used in this research from the Africa-based crime dataset [9]. These data are in Comma Separated Value (CSV) and are passed into the data pre-processing stage for further processes, see Fig. 1.

3.2 Stage 2: Data Pre-processing (ETL Process)

The acquired data is pre-processed to make the data suitable for the experiments and remove the outliers or noises that can affect the performance of the proposed model. In this research, the data is pre-processed through the concept of Extraction, Transformation and Loading (ETL) [10]. The raw data is extracted from different sources and is passed into the data transformation stage where the raw data is converted into a meaningful dataset. This is then loaded to the feature selection stage for further processes.

3.2.1 Feature Selection and Periodic Prediction Features

Here the pre-processed data is fed as input to the feature selection component where the relevant features from the original dataset are selected, and irrelevant features are removed. The crime features used in this research are the baseline and the derived

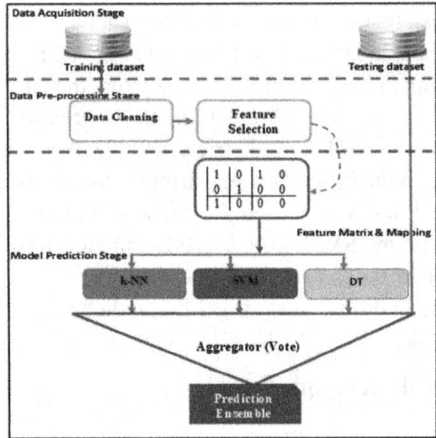

Fig. 1. System framework for prediction of possible crime locations.

features. Crime feature baselines are those features that are obtained directly from the original crime dataset; these attributes are: date, crime description, crime resolution, crime type, crime location, address, X-coordinate,and Y- coordinate. However, security operatives and sociologists have argued that the week periods, time periods and festive periods contribute to environmental crime types and rates [11, 12]. The information in this incidence [13] testifies to variability in crime rates due to periods, week periods, and festive periods. These features have been used by security operatives and sociologists in other research work [14]. Hence, to improve the prediction performance, the new features are derived and defined as the periodic features of the original datasets. These features are:

- Period: the week the crime is committed. For instance, weekdays = Monday – Thursday, weekends = Friday – Sunday.
- Time: the specific time of the day (i.e., during the day or night) at which the crime is being committed.
- Festive: this is the festive or celebration time within the year in which crimes are committed.
- Population: this is the number of people living in the geographical location where the crime is committed.

3.3 Stage 3: Modelling and Prediction

3.3.1 Model Training and Testing

In this stage, the ensemble method based on emerged models (KNN, SVM, and DT) are trained on the crime dataset sample. After the training, the performance of the model is checked based on accuracy and errors. Then the trained model is tested on the unseen dataset, which is done by using the cross-validation technique, which involves the splitting of the trained data to 90% for training and 10% of testing data is used to test the model.

3.3.2 Building Model by Ensemble Methods

Ensemble learning uses multiple machine learning models to make better predictions on a dataset. An ensemble model works by training different models on a dataset and having each model make predictions individually. The predictions of these models is aggregated and the majority vote is used to make final prediction.

3.4 Evaluation Metrics

To evaluate the performance of the new model, the confusion matrix and hold-out cross-validation technique are adopted, the detailed metrics can be found in [15].

4 Experimental Evaluations and Results

4.1 Data Description and Experimental Settings

The implementation software used in this research is Python. The crime dataset used is obtained from the Africa crime database repository [9].

Africa-based Dataset:
The South Africa (SA) crime dataset used in this implementation starts from April 2015 to June 2019 [9]. These data consist of over 339471 records with 6 variables. As part of the contributions of this research, more attributes are derived as periodic to the crime dataset to improve the crime link prediction performance of the proposed ensemble method. The link features are obtained when there are connections between the crime type that occurred on a specific day and time of the year to a location, then, such link is manually labelled as '1'. However, if the crime type that occurs on a particular day and time of the year has no connections to a location, then it is labelled as '0'. The samples of a few Africa-based crime datasets used for the implementations, see Table. 1.

Table 1. Samples of the Africa-based crime dataset.

Days	Period	Time	Festive	Population	Crime Type	Crime Location	Link
Sunday	weekend	Day	No	4,618,741	Theft	Cape Town	1
Sunday	weekend	Day	No	5,635,137	Theft	Johannesburg	0
Sunday	weekend	Day	No	3,721,628	Theft	Durban	0
Wednesday	weekday	Night	No	2,473,000	Assault	Pretoria	0
Tuesday	weekday	Day	No	4,618,741	Burglary	Cape Town	1
Tuesday	weekday	Day	No	3,721,628	Burglary	Durban	0

(*continued*)

Table 1. (*continued*)

Days	Period	Time	Festive	Population	Crime Type	Crime Location	Link
Wednesday	weekday	Day	No	5,635,137	Vandalism	Johannesburg	0
Thursday	weekday	Night	No	267,007	Theft	East London	0
Wednesday	weekday	Day	No	110,000	Theft	Mbombela	0
Friday	weekday	Night	No	4,618,741	Assault	Cape Town	1
Friday	weekend	Day	No	130,000	Assault	Polokwane	0
Friday	weekend	Day	No	5,635,137	Vandalism	Johannesburg	0

Legend: Weekday = Monday - Thursday; Weekend = Friday - Sunday

4.2   Experiment 1: Predicting Future Crime Locations from Africa-Based Data

This research aim is to bring the theory of different classifiers to practice with application to the prediction of the future link between crime type that occurs on a specific day and time of the year to a location with improved results. To justify the improved results of the classifiers, an Africa-based crime dataset is used [9, 16]. Different experiments were conducted on the datasets, and the predicted performance results of different classifiers were obtained using cross-validation, Receiver Operating Characteristics (ROC) curve, and confusion matrix described in [17]. The various experiments conducted are explained in detail in subsequent sections.

4.2.1   Experiment 1.1 Predicting Future Crime Locations Without Periodic Features

Our intention here is to determine whether the proposed model can accurately predict the likely future link between crime locations and crime types that occurs on a specific date of the year without periodic features. For instance, the security operatives may ask: what would be the likelihood of a future link occurrence between a specific crime type that occurs on a specific date to a location on a particular day of the week?

$$\{(\text{Link?}|\text{Location} = \text{Pretoria, Crime Type} = \text{theft, Date} = 04\text{-}03) = ?\}$$
$$\{(\text{Link?}|\text{Location} = \text{Johannesburg, Crime Type} = \text{murder, Date} = 23\text{-}04) = ?\}$$
$$\{(\text{Link?}|\text{Location} = \text{Cape Town, Crime Type} = \text{burglary, Date} = 23\text{-}12) = ?\}$$
$$\{(\text{Link?}|\text{Location} = \text{Pretoria, Crime Type} = \text{Theft, Date} = 25\text{-}12) = ?\}$$

$$(1)$$

In the first scenario of Eq. (1), the security operator wants to know the most probable future link when the location is Pretoria and the theft crime type on 04-03, and this is similarly repeated for the rest of the situations in Eq. (1). Hence, the revealed states of the knowledge of all the linking scenarios in Eq. (1) are therefore shown to understand the future relationship between crime type that occurs on a specific date and time of the year to a location using the historical crime dataset. A cross-validation technique was therefore used in the evaluation, where 90% of the dataset was used for training and 10% was used for evaluation. This experiment was repeatedly done on Logistic Regression Model (LRM), Random Forest (RF), Naïve Bayes (NB), Support Vector Machine (SVM), k-Nearest Neighbor (k-NN), and Decision Tree (DT). The prediction results of each experiment, see columns 5 to 10 of Table 2.

Table 2. Evaluation results of future crime location prediction without periodic features.

Date	Crime Type	Crime Location	Target Link	Predicted					
				LRM	RF	NB	SVM	k−NN	DT
04-03	Murder	Cape Town	1	0	1	0	1	1	1
17-03	Theft	Pretoria	0	0	1	0	0	0	0
14-04	Theft	Durban	0	1	1	0	1	0	0
23-04	Murder	Johannesburg	0	1	0	0	0	1	0
14-05	Murder	Cape Town	1	1	1	1	1	1	1
23-05	Burglary	Pietermaritzburg	1	0	0	1	0	0	1
07-06	Murder	Pretoria	1	1	0	1	1	0	0
20-06	Murder	Pretoria	1	0	0	0	0	0	1
26-06	Theft	Mbombela	0	0	1	0	0	0	0
23-12	Burglary	Cape Town	1	0	1	1	1	0	1
25-12	Theft	Pretoria	1	1	0	1	1	1	1
31-12	Murder	Johannesburg	1	1	1	0	1	1	1

Table 2 presents the evaluation result of the six classifiers with their capability of understanding the likely future link between crime locations and crime types that occurs on a specific date and time of the year ahead, using an Africa-based crime dataset. The predicted results have therefore emerged from the evaluation data and the actual crime results are revealed from the training dataset as shown in Table 2. The confusion matrix for each classifier is obtained as shown in Fig. 2.

From Fig. 2, the diagonal of the confusion matrix represents the number of correct predictions in each crime location category. To obtain a more complimentary view of the results of the classifiers, the ROC curves are established, see Fig. 3.

Figure 3 shows the ROC curves which represent the relative trade-offs between True Positive Rates (TPRs) and False Positive Rates (FPRs) of each classifier used in the experiment for the prediction of the possible future link between crime type that occurs on a specific day of the year to a location. From Fig. 3, the correctly predicted links are computed as TPRs, while the incorrectly predicted crime link or false alarms are the FPRs. From Fig. 3, one can see that the TPRs of the k-NN, SVM, and DT are higher while their FPRs are lower. This shows that the three models could be used as an ensemble model for efficiently future crime link prediction. Furthermore, other metrics such as

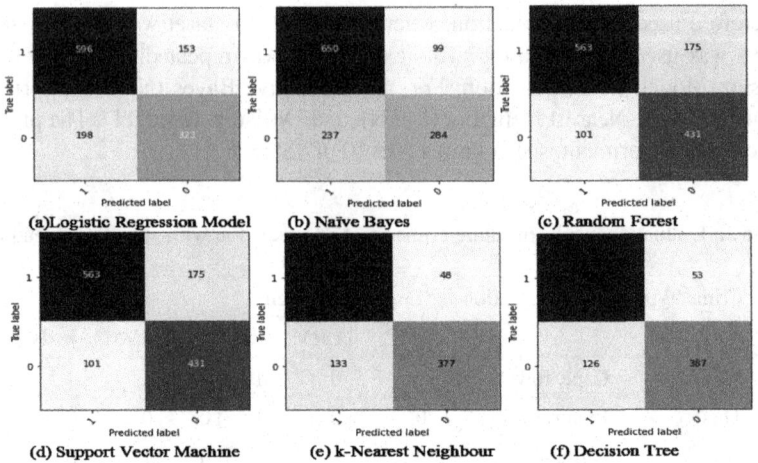

Fig. 2. Confusion matrices capturing the performances of the implemented models.

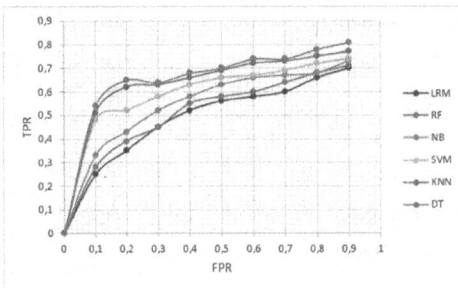

Fig. 3. ROC-curve of six classifiers used in the experiment on a few crime features on an Africa-based crime dataset.

precision, recall, F1-score, and accuracy of all the six models used are evaluated and the summary of the overall implementation performance, see Table 3.

The performance comparison of different classifiers used for the prediction of the next crime locations in this research, see Table 3. To obtain the best three classifiers as well as to further improve the classifier performances, the voting method is adopted. The basic idea of voting is to use multiple classifiers trained on the dataset and then decide on the best three classifiers based on the majority votes. The k-NN, SVM, and DT (ensemble model) give better results with an average precision of 0.90, recall of 0.91, F1-score of 0.91, and accuracy of 92.24%. These results are superior to the other models used during the implementation of the future link prediction between the crime locations and crime types that occur on a specific date and time of the year.

Table 3. Performance metrics.

Models	Performance			
	Precision	Recall	F1-score	Accuracy (%)
LRM	0.74	0.72	0.73	74.41
NB	0.79	0.77	0.78	78.94
RF	0.81	0.80	0.81	80.27
k-NN	0.89	0.90	0.89	89.75
SVM	0.90	0.92	0.91	91.05
DT	0.92	0.90	0.93	92.91
Average of proposed Ensemble method (k-NN + SVM + DT)	0.90	0.91	0.91	92.24

Reasoning on Specific Scenarios.
The security operatives may now use the proposed ensemble model to answer the below queries from the implementation results as shown in Eq. (2).

$$\{(\text{Link?}|\text{Location} = \text{Pretoria, Crime Type} = \text{theft, Date} = 04\text{-}03) = 1\}$$
$$\{(\text{Link?}|\text{Location} = \text{Johannesburg, Crime Type} = \text{murder, Date} = 23\text{-}04) = 0\}$$
$$\{(\text{Link?}|\text{Location} = \text{Cape Town, Crime Type} = \text{burglary, Date} = 23\text{-}12) = 1\}$$
$$\{(\text{Link?}|\text{Location} = \text{Pretoria, Crime Type} = \text{Theft, Date} = 25\text{-}12) = 1\}$$

$$(2)$$

From the security analysis scenarios in Eq. (2), the security operators might want to know the possibility of a link between crime locations and crime types at a specific date of the week. This gives security analysts insight into the link between crime locations and types to minimize crime from occurring. Similarly, answering the three guideline questions of the crime link that follows Table 2 using the prediction knowledge revealed by the proposed ensemble model, over days of the week, facilitates better anticipatory security planning and making an astute decision.

Q1: What is happening?
A1: In the scenarios, it is evident that the theft crime type is likely to occur in Pretoria on 04-03. This might be due to social factors such as the populations of the area, the period of the day, etc. Furthermore, in Cape Town and Pretoria on 23-12 and 25-12, there is a link between a crime location and burglary and theft crime type as shown in Table 2. This might be because of the population of people at this location on these specific days which are also the festive period in the year. However, these prediction results are not similar (inaccurate) to the original target link in Table 2 due to the few data features used during the prediction.

Q2: Why is it happening?
A2: The crime rate is dominated the weekdays in highly populated locations as shown

in Table 2. This shows the link of crime type with the location has a sharp response to the weekdays, highly populated areas, and festive period. Although the proposed ensemble method could not accurately predict the link when compared with the original link target as illustrated in Table 2. The inaccuracy might be a result of fewer attributes used for the implementation.

Q3: What can the security operators do?
A3: Concerning the information obtained from Table 2, in highly populated locations, on weekdays, as well as festive days of the year, the security operators must deploy their resources and pay full attention to these locations to minimize the reoccurrence of the crime.

4.2.2   Experiment 1.2: Predicting Future Crime Locations with Periodic Features

The objective of this experiment is to determine whether adding periodic features to the crime dataset can improve the crime link prediction model results. The security operatives may ask a similar question as in Experiment 1.1. The experiment is conducted like that in Experiment 1.1 on the six selected classifiers. For the predicted results, see columns 9th to 14th of Table 4.

Table 4. Results of future crime location prediction with periodic features.

Days	Period	Time	Festive	Population	Crime Type	Crime Location	Target Link	Predicted					
								LRM	RF	NB	SVM	k-NN	DT
Monday	weekend	Day	No	4,618,741	Murder	Cape Town	1	0	1	1	1	1	1
Sunday	weekend	Day	No	2,473,000	Theft	Pretoria	0	0	1	0	0	0	0
Sunday	weekend	Day	No	3,721,628	Theft	Durban	0	1	1	0	0	0	0
Wednesday	weekday	Night	No	5,635,137	Assault	Johannesburg	0	1	1	1	0	1	0
Tuesday	weekday	Day	No	4,618,741	Murder	Cape Town	1	1	1	0	1	1	1
Tuesday	weekday	Day	No	130,000	Burglary	Pietermaritzburg	0	0	0	1	0	0	0
Monday	weekday	Day	No	2,473,000	Murder	Pretoria	1	1	0	1	1	1	1
Thursday	weekday	Night	No	2,473,000	Murder	Pretoria	1	1	0	0	1	1	1
Wednesday	weekday	Day	No	110,000	Theft	Mbombela	0	1	0	1	0	1	0
Monday	weekday	Night	Yes	4,618,741	Burglary	Cape Town	1	0	1	1	0	1	1
Wednesday	weekend	Day	Yes	2,473,000	Theft	Pretoria	1	0	1	1	1	0	1
Tuesday	weekend	Day	Yes	5,635,137	Murder	Johannesburg	1	1	1	0	0	1	1

Table 4 presents the evaluation results of the different classifiers used in the experiments with their capability of understanding the future link between the crime type that occurs on a specific day and time of the year to the locations using the historical crime locations obtained from Africa-based crime datasets. To be more precise, a confusion matrix is utilized for each classifier, see Fig. 4.

To further validate the performance of the classifiers used in the experiment based on the Africa-based crime dataset, the ROC curves for each classifier are established, see Fig. 5.

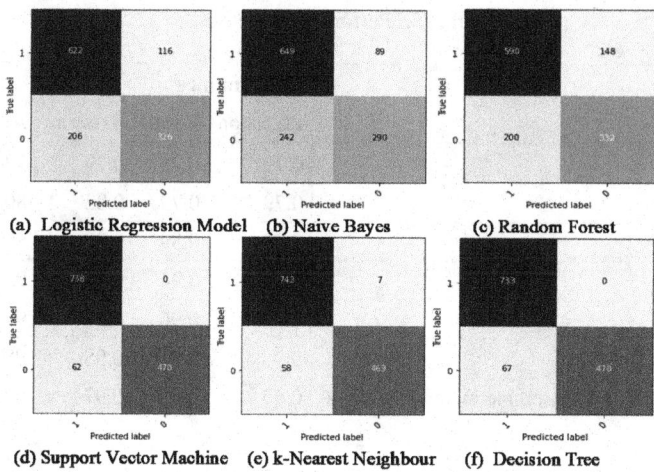

Fig. 4. Confusion Matrices capturing the performances of the implemented models.

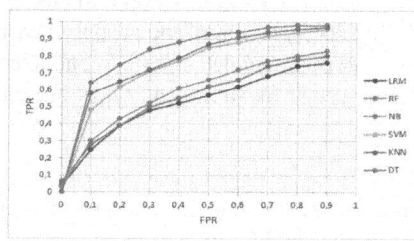

Fig. 5. ROC-curve of six classifiers used in the experiment on a derived crime features of an Africa-based crime dataset.

A close investigation of Fig. 5 reveals an interesting result regarding the performance of different classifiers used. It is important to know that the higher the true positive rate value the better the model. In comparison, one can see from the ROC curve that some algorithms have higher TPRs than others for the predictive models when used for predicting the likely future link between crime locations and crime types on a specific day of the year. The summary of the overall implementation performance in terms of precision, recall, F1-score, and accuracy, see Table 5.

In Table 5, the performance comparison of different classifiers used for the prediction of the likely link between future crime locations and types in this research is shown with improved performance results compared to that of Experiment 1.1. This improved prediction result is because of the more crime attributes used on the training dataset during the experiment. To obtain the best three classifiers with good predictive results, a voting technique is adopted, and the best three classifiers are decided by a majority vote are the k-NN, SVM, and DT models which give accuracies of 91.65%, 92.23%, and 98.75% respectively. The overall performance of these three techniques gives a precision of 0.97, recall of 0.95, F1 score of 0.97 and accuracy of 97.03%. One can observe that the proposed ensemble model gives improved crime link prediction results compared to

Table 5. Performance metrics.

Models	Performance			
	Precision	Recall	F1-score	Accuracy (%)
LRM	0.79	0.78	0.79	79.65
NB	0.79	0.79	0.82	80.59
RF	0.89	0.89	0.88	88.59
k-NN	0.95	0.94	0.95	95.12
SVM	0.97	0.96	0.97	97.23
DT	0.98	0.96	0.98	98.75
Average of proposed Ensemble method (k-NN + SVM + DT)	0.97	0.95	0.97	97.03

other selected models, see Table 5. This improved model is due to the combination of three different classification models to form the proposed ensemble model.

For the overall performance comparison of the proposed ensemble model, we summarized in Table 6, the two experimental test cases conducted on the prediction of the likely link between future crime locations and types in terms of precision, recall, F1-score, and accuracy.

Table 6. Comparing performance analysis of link prediction for experimental test cases.

Experimental Test Cases	Precision	Recall	F1-Score	Accuracy (%)
Experiment 1.1	0.90	0.91	0.91	92.24
Experiment 1.2	0.97	0.95	0.97	97.03

From Table 6, it can be observed that the precision, recall, F1-score, and accuracy of 0.90, 0.91, 0.91, and 92.24%, respectively were achieved in Experiment 1.1 which was on the proposed model with fewer datasets while Experiment 1.2 which utilizes more features on the proposed ensemble achieved a precision of 0.97, recall of 0.965 F1-score of 0.97, and accuracy of 97.03%. One can observe clearly from these results that the proposed ensemble model achieved improved performances in Experiment 1.2 compared to Experiment 1.1 because of the many derived features that were added to the baseline features. This result attests to the inference made in [18] that more derived dataset features can improve system predictive performances. Furthermore, with the improved performance of the new model, this model can successfully be used as an application to augment the work of security personnel in the prediction of a future link between the crime locations and crime types, and this will ensure effective allocation of security resources to appropriate locations before the crime occurs.

Reasoning on Specific Scenarios

Hence, the security operatives may now use the proposed ensemble model to answer the below queries from the implementation results as in Eq. (3).

{(Link? | Location = PTA, Crime Type = theft, Population = high, Day = Mondays) = 1}.

{{(Link? | Location = JHB, Crime Type = murder, Population = high, Day = Tuesdays) = 1}.

{(Link? | Location = Cape Town, Crime Type = burglary, Population = low, Day = Mondays) = 1}.

{(Link? | Location = Pretoria, Crime Type = Theft, Population = low, Day = Wednesdays) = 1}

$$(3)$$

Having seen the crime link prediction over time in Eq. (3), the security operators can decide after knowing what was not known by answering the four questions on security about crime link locations. Detailed understanding and decision-making process of the scenarios are shown in Table 4.

Q1: What is happening?
Q1: What is happening?

A1: In Table 4, when comparing the link prediction results of the proposed ensemble model between theft crime types at the Pretoria location in the first scenario with the actual target link, one can see that there are similarities in the link. This shows that there is a correlation between the crime type that occurs on Mondays in Pretoria where there is a high population of people. Also, there is a link between a crime that occurred in Cape Town and Pretoria with a high population of people on Mondays and Wednesdays.

Q2: Why is it happening?
A2: From Table 4, it is evident that in Pretoria and Cape Town there is the possibility of the burglary and theft crime types reoccur. This is because the periods are a festive time and people are everywhere preparing and planning for the celebration. Hence, criminals take advantage of such periods to manifest their criminal acts. From Table 4, one can see that the predictive model link prediction is consistent with the original target link.

Q3: What can the security operators do?
A3: It is a clear indication that the security operators must ensure that they pay close attention and allocate their resources to Pretoria and Johannesburg, especially during weekdays, to avert crime from reoccurring at these locations.

5 Conclusion and Future Work

In this section, the ensemble model based on (DT, SVM and k-NN) has been developed and presented as a new crime link prediction model to predict the link between the crime locations and crime types on a specific day of the year. The proposed model is rigorously subjected to several prediction evaluations using Africa -based crime dataset which is obtained from South Africa Police Services (SAPS) data repository. Since the

DT, SVM, and k-NN have sufficiently shown improved link prediction performances during implementations, the security operators now have fewer worries to decide the future link between the crime type on a specific day of the year to the location in real-life applications.

We have methodologically presented different experimental results obtained on the baseline link predictive models and the proposed ensemble model for future link prediction of crime type that occurs on a specific day and time of the year to a location using locally obtained crime dataset without periodic features produced an overall average precision, recall, F1-score, and accuracy of 0.90, 0.91, 0.91, and 92.24%, respectively, see Table 2. When tested on the crime dataset with periodic features the proposed model produced an overall average precision, F1- score and accuracy are 0.97, 0.97, and 97.03% respectively, see Table 5. These results accuracies are more accurate and superior when compared with accuracies of other selected predictive models used as the baseline for predictive comparison by other researchers. These results have proven that the proposed ensemble model can accurately predict in the future the link between crime types at specific days and period of the year to locations when compared with other predictive models.

This study has demonstrated that factors such as period, the population of people, and the festive period can contribute to crime in a particular location by conducting different experiments using the proposed ensemble model, and results have shown the performance of the proposed model. Hence, utilizing the proposed model could assist security operators to fight crime effectively and efficiently by allocating their resources to the locations before the crime occurs. In future, work could be done on the visualization of the crime links to reveal possible hidden criminal locations to the security operatives.

Acknowledgements. The authors acknowledge the financial support made available by the University of South Africa and resources made available by Norfolk University, USA.

References

1. Esan, O.A., Osunmakinde, I.O.: Towards intelligence vision surveillance for police information systems. In: Silhavy, R. (eds) Cybernetics Perspectives in Systems. CSOC 2022. Lecture Notes in Networks and Systems, vol. 503 (2022). https://doi.org/10.1007/978-3-031-09073-8_13
2. Felix Enigo, V.S.: An automated system for crime investigation using conventional and machine learning approach. In: Raj, J.S., Bashar, A., Ramson, S.R.J. (eds.) ICIDCA 2019. LNDECT, vol. 46, pp. 109–117. Springer, Cham (2020). https://doi.org/10.1007/978-3-030-38040-3_12
3. Belesiotis, A., Papadakis, G., Skoutas, D.: Analyzing and predicting spatial crime distribution using crowdsourced and open data. ACM Trans. Spatial Algorithms Syst., **3**(4), 12:1–12:31 (2018)
4. Esan, O.A., Osunmakinde, I.O.: A computer vision model for detecting suspicious behaviour from multiple cameras in crime hotspots using convolutional neural networks. In: Highlights in Practical Applications of Agents, Multi-Agent Systems, and Complex Systems Simulation. The PAAMS Collection. PAAMS 2022. Communications in Computer and Information Science, vol. 1678 (2022). https://doi.org/10.1007/978-3-031-18697-4_16

5. Khan, M., Ali, A., Alharbi, Y.: Predicting and preventing crime: a crime prediction model using san francisco crime data by classification techniques. Wiley Hindawi **22**, 1–13 (2022). https://doi.org/10.1155/2022/4830411

6. Stalidis, P., Semertzidis, T., Daras, P.: Examining deep learning architectures for crime classification and prediction. Forecasting **3**, 741–762 (2021). https://doi.org/10.3390/forecast3 040046

7. Castro, U.R.M., Rodrigues, M.W., Brandao, W.C.: Predicting crimes by exploring supervised learning on heterogenous data. In: Proceedings of the 22nd International Conference on Enterprise Information Systems, vol. 1, pp. 524–531 (2020)

8. Rajadevi, R., Devi, E.M.R., Kumar, S.V.: Prediction of crime occurrence using multinomial logistic regression. Int. J. Innov. Tech. Exploring Eng. (IJITEE) **3**(3), 1432–1435 (2020). https://doi.org/10.35940/ijitee.B7663.019320

9. S. Wessels, South Africa Crime Dataset (2017). https://www.kaggle.com/slwessels/crime.sta tistics-for-South-Africa

10. Yerpude, P., Gudur, V.: Predictive modelling of crime dataset using data mining. Int. J. Data Mining Knowl. Manag. Process (IJDKP) **7**(4), 43–58 (2017)

11. Garton, N., Niemi, J.: Multivariate temporal modeling of crime with dynamic linear models. PLOS ONE **14**(7), (2019). https://doi.org/10.1371/journal.pone.0218375

12. Jacob, B., Lefgren, L., Moretti, E.: The dynamic of criminal behaviour, evidence from weather shocks. J. Hum. Resour. **42**(3), 489–527 (2007)

13. Towers, S., Chen, S., Malik, A., Ebert, D.: Factors influencing temporal patterns in crime in a large American city: a predictive analytics perspective. PLoS ONE **13**(10), (2018). https://doi.org/10.1371/journal. pone.0205151

14. Rumi, S.K., Deng, K., Salim, F.D.: Crime event prediction with dynamic features. EPS Data Sci. **7**(43), 1–27 (2018)

15. Yu, T., Yan, J., Lu, W.: Combining background subtraction and convolutional neural network for anomaly detection in pumping-unit surveillance. Algorithm **12**(115), 1–13 (2019). https://doi.org/10.3390/a12060115

16. S.F.C. dataset, San Francisco Crime Statistics 2005–20015 (2020). https://sfgov.org/crime-statistics

17. Esan, O.A., Osunmakinde, I.O.: Towards intelligent vision surveillance for police information systems. In: CSOC, Lecturer notes in Networks and Systems, vol. 503 (2022)

18. Borkin, D., Nemeth, M., Michalconok, G., Mezentseva, O.: Adding additional features to improve time series prediction. Res. Papers Faculty Mater. Sci. Technol. **27**(45), 72–78 (2019)

An Empirical Analysis on Lossless Compression Techniques

Mohammad Badrul Hossain$^{(\boxtimes)}$ and Md. Nowroz Junaed Rahman

BRAC University, 66 Mohakhali, Dhaka, Bangladesh
{badrul.elprincipe22,nowrozjunaedrahman}@gmail.com

Abstract. A method of presenting source data into it's compact form is known as data compression. In this process, data size is minimized, redundancy is eliminated, and excess information is gotten rid of. A reduction in actual data is usually advantageous because it uses less resources overall, including bandwidth, processing, space, time, and many others. There are numerous compression algorithms for reducing the size of data of different formats. Even for compressing a particular data type, many approaches are being used. The proposed research has explored three of the lossless compression techniques which are: Run Length Encoding, Lempel Ziv Welch, and Huffman Encoding algorithms. We found out that based on compression size, compression ratio, and space saving percentage, Lempel Ziv Welch outperformed the other two. In contrast, Huffman Encoding performed better than the other two based on compression time. In the best case, LZW got a compression size of 250992 bytes, a compression ratio of 5.0106, and a space saving percentage of 80.04% while Huffman encoding got a compression time of 32.28 ms.

Keywords: Run Length Encoding · RLE · Dictionary · Lempel Ziv Welch · LZW · Huffman · compression · text compression · text-file · lossless compression

1 Introduction

A short while ago, transferring a stream of data was slow and inconvenient, not to mention very inefficient. After the introduction of compression techniques in computer applications, a new dimension in digital world has been created. More than 2.5 quintillion bytes of data are produced and collected by humans every day [1]. Over 90% of the data generated in the span of human history has been created in the last few years, and it was estimated that it would exceed 40 trillion gigabytes by 2020 [2]. Therefore, considerably more advanced and effective compression methods will be required in near future.

The main function of compression is that it makes data smaller. This is incredibly advantageous when transferring a lot of data. After using the technique, there should be a significant difference between the source and compressed

© The Author(s), under exclusive license to Springer Nature Switzerland AG 2023
F. Neri et al. (Eds.): CCCE 2023, CCIS 1823, pp. 158–170, 2023.
https://doi.org/10.1007/978-3-031-35299-7_13

data. Main goal of compression in data transmission applications is to increase speed. On the other hand, the degree of compression is the primary concern for storage devices [3]. There can be a couple of compression types, which are lossless and lossy. When lossless compression is used, the original data can be restored from the compressed data without losing any information. Lossless compression is responsible for preserving computer executable files, medical images, and for many other important reasons [4]. On the contrary, the lossy compression method results in some data loss upon decompression. Therefore, it is not possible to get back the original data. Although lossy compression techniques fail to revert back to source data after decompression.

The major contributions of our research are as follows: firstly, theoretical analysis on RLE, LZW and Huffman encoding was done to understand their diverse and distinct nature during data compression. Secondly, these algorithms were implemented using Java and they were used for compressing several files containing text data from Canterbury corpus [5]. Thirdly, we analyzed the performances of these algorithms based on compression size, compression ratio, space saving and compression time. From the analysis the findings are, LZW shows the most efficiency in terms of compression size, ratio, and space saving. However, Huffman encoding performs best in compression time.

The paper is organized as follows: Sect. 2 contains some related works that have been done on this topic. A technical overview of RLE, LZW and Huffman Encoding has been given in Sect. 3. Section 4 will discuss the implementations of the above-mentioned algorithms. In Sect. 5, empirical analysis of the implemented algorithms will be performed and illustrate the results based on different metrics. Lastly, Sect. 6 will conclude the analysis by deciding which compression approach performs better.

2   Related Works

To increase the efficiency of different compression techniques, a lot of research is going on. Sometimes it is seen that, while compressing the data with dictionary, the dictionary size gets significantly larger compared to the actual data. Therefore, Brisaboa et al. [6] proposed a way to even shrink the dictionary for faster service in his paper. Brisaboa et al. [7] had done a great job recently by using the longest common prefix. They designed a new data structure for achieving faster storage and query time.

In their paper, Kempa et al. [8] showed how their universal data structure works for all the types of dictionary compression algorithms which also supports random access in any dictionary scheme.

Nishimoto et al. [9] demonstrated how they employed a practical bidirectional parsing in LZ77 compression algorithm which is a dictionary based compression algorithm. In this work, the authors proposed a practical bidirectional parsing called *right reference* which guarantees to generate less number of phrases compared to the original LZ77.

In their research, Platoš et al. [10] performed compression by taking sequences of words rather than sequences of bytes or characters. They employed three compression algorithms in their work, which were Huffman encoding, LZ77, and LZW. Among these three, LZ77 outperformed the others.

Javed et al. [11] worked on constructing an automatic page segmentation that functioned by taking compressed text documents from Run Length Algorithm. They experimented on 166 types of compressed documents from MARG dataset and acquired an accuracy of 95%.

In their work, Fiergolla et al. [12] proposed a set of pre-processing steps to greatly improve the performance of RLE. As per their proposal, feeding input data into byte-wise encoding will generate a bit string but a bit string does not guarantee a long run of repeated characters so, this is why they proposed to arrange bits in a specific manner. They suggested reading all significant bits first then, reading the second most significant bits, and so on and so forth. This way the longer average run length of repeating characters is ensured.

3   Preliminaries

Before starting the analysis glimpse of the workflows of Run Length Encoding, LZW Compression, and Huffman Encoding algorithms will be given.

3.1   Run Length Encoding

Run Length Encoding is one of the most primitive compression algorithms which is basically functional only for repeated characters. For example, a string containing characters $aaaabbbdddaa$ of 12 bytes can be compressed to $a4b3d3a2$ character sequence by using RLE approach which requires 8 bytes. In the worst-case scenario, the compressed data might be twice the size of the actual data.

3.2   LZW Compression Algorithm

LZW is a form of dictionary based lossless compression algorithm. Dictionary compression techniques rely upon the observation that there are correlations between parts of data. The basic idea is to replace those repetitions with references to a "dictionary" containing the original [13]. In order to get rid of anomalies caused for redundancy characteristics during compression, LZW was proposed by Terry A. Welch in 1984 which is an adaptive version of the previously developed Lampel Ziv Algorithm [14]. This algorithm starts functioning by storing all the ASCII codes. Then if a new character sequence gets found which is not stored earlier it will be stored with the previously stored characters by assigning a new ASCII value. The entire process continues until a given string gets fully parsed. For example, for string $AAABBBZ$, default ASCII values from 0–255 are already stored in the dictionary where A, B and Z have the values of 65, 66, and 90 respectively. During reading the given string firstly A will be found which is already stored in the dictionary but for the next two characters AA no assigned ASCII value exists in the dictionary. Therefore, a new ASCII value will be assigned to the character sequence AA which is the very next value of the lastly assigned value 255. That means 256 will be assigned to *AA* and added to the dictionary. Then *B* will be found which already exists and it will continue the process. Afterward, for the new sequence *BB* it will generate the value 257 and add it. Lastly, *Z* will be read and as it exists in the dictionary, the entire string will be compressed according to the values stored in the dictionary. For the above mentioned string compressed data will be, [65, 256, 66, 257, 90]

3.3 Huffman Encoding

Huffman coding was proposed by David A. Huffman in 1952 which generates binary code for each character according to their occurrences for encoding [15]. In Huffman Encoding, each character gets sorted in ascending order according to the corresponding character frequencies and enlisted as terminal nodes of a binary tree. Two nodes containing the lowest frequencies generate a parent node for them containing the sum of the frequencies of those two nodes. The frequency of the parent node will be considered as the new frequency of the tree and the newly generated tree will be placed in a new place on the list maintaining the ascending order of the frequencies. The entire process continues until only one binary tree gets constructed, which is known as Huffman Tree. Huffman Tree generates 0 for traversing any of its left child nodes and 1 for traversing right child nodes. That is how codewords for the characters are generated and encoded data represents separate binary codewords for each character.

4 Implementation

For the analysis, all three algorithms were implemented in Java using *NetBeans* IDE. Objects from the *String class* were required for the Run Length approach. *Collection class* objects from java built-in libraries were needed for the implementation of LZW and Huffman alongside the *String class*. *System.nanoTime()* function was used for calculating the compression time for each approach. In order to compare the sizes of data before and after compression, the number of characters was multiplied by 8 each time before and after the compression for all three algorithms.

4.1 Run Length Encoding

Implementation of RLA was done by importing the String class of java. In Algorithm 1, function *Encode()* is basically taking String data from text files as input, compresses the data into the encoded string, and returns it.

Algorithm 1. RLE

Input: Text data
Output: Compressed data
function *Encode(str)*:
    for: *i=0 to length of str-1*
        count ← 1
        while: *i is **less** than length of str-1 **And** str.charAt(i) **equal** to str.charAt(i+1)*
            count = count+1
            i = i+1
        s ← s+*str*.charAt(i)+count
    return s
end function

4.2 LZW Compression Algorithm

Algorithm 2. LZW

Input: Text data
Output: Compressed data
function *Encode(str)*:
    size ← 256
    for: *i=0 to size-1*
        dict.put((char)i, i)
    for: *char c : str.toCharArray()*
        wc ← w+c
        if: *dict.containsKey(wc) is **True***
            w ← wc
        else:
            result.add(dict.get(w))
            dict.put(wc,size++)
            toShow.put(wc,size-1)
            w ← " "+c
    if: *w is **Not** an empty string*
        result.add(dict.get(w))
    return result
end function

For LZW two separate *Map* objects consisting of the same parameters were used in Algorithm 2. *Map* named *dict* is being used for storing ASCII values in the dictionary and values for newly generated character sequences are being stored in *toShow*. Both of them are using *string* data type as *keys* and *integers* for representing ASCII codes. The *Encode()* function takes *string* as input and then compresses the data by storing the codes in the above mentioned maps. Finally, the encoded data is stored in an *arraylist* named *result*. After the completion of encoding the *arraylist* gets returned as output.

4.3 Huffman Encoding

Algorithm 3. Huffman Encoding

Input: Text data
Output: Compressed data
function *getCharFreq(str)*:
   for: i=0 to length of str-1
      ch ← str.charAt(i)
      if: map.containsKey(i) is **Not True**
         map.put(ch,1)
      else:
         val ← map.get(ch)
         map.put(ch,++val)
   return map
end function
function *encode(*charCode,*str)*:
   initializes a *StringBuilder*: sb
   for: i=0 to length of str-1
      sb.append(charCode.get(*str*.charAt(i)))
   return sb.toString()
end function
function *compress(str)*:
   charFreq ← getCharFreq(*str*)
   root ← buildTree(charFreq)
   charCode ← createHuffmanCodes(charFreq.keySet(),root)
   encodedMsg ← encode(charCode,*str*)
end function

According to Algorithm 3, the procedure of compression starts from the *compress()* function which takes *string* input. Map *charFreq* consisting of two parameters *string* for key and *integer* for frequency stores the information of frequencies of each character which is basically returned from the invocation of the function *getCharFreq()*. This function has the only parameter which takes the input *string*. A *map* is being used for generating the frequencies for each character in this function and it gets returned. *HuffmanNode* of four parameters named *root* is being generated by invoking the *buildTree()* function which takes the *map charFreq* as input. The function generates a priority queue according to the frequencies of the characters extracted from the *map* and then constructs the tree by creating a parent node by combining the characters consisting of the lowest frequencies. For *HuffmanNode* there are four parameters; namely the character, *integer* for frequency, references of left child and right child respectively. In the *map charCode* having the same parameters as the previous *maps* separate Huffman Codes for each character is being stored. This is the consequence of the execution of the *createHuffmanCodes()* function. The function executes by taking two parameters *charFreq map* and *root* Huffman

Tree. It then generates the codes by traversing the tree. If traversal gets through the left child, 0 is generated and for the right child, it generates 1. This is how the function creates binary codes for each character and stores them in a *map* then returns it. Finally, the encoded message is stored in *encodedMsg* which is a *string* data. The final phase of compression gets done by the execution of the *encode*() function which takes *charCode map* and the input data *str* as inputs. Inside this method, a *StringBuilder* is being used in order to construct the encoded data in *string* format.

5   Comparison and Result Analysis

In this section, we will compare the performance of each of the three compression algorithms based on the following factors: compression time, compression ratio, space-saving, and compression size. To evaluate and compare the performance of these algorithms, they were applied to a collection of text files from the Canterbury Corpus [5].

5.1   Measurement Factors

It is already been mentioned which factors were considered to measure the performance of the algorithms. In this sub-section, we will try to give a brief idea about these factors.

Compression Ratio: Compression ratio indicates the ratio between the actual size of the text file and the size of the text file after compression.

$$\frac{Original\ Fize\ Size}{File\ size\ after\ compression} \tag{1}$$

Saving Percentage: Saving Percentage indicates the amount of file size reduced after compression in percentage.

$$1 - \frac{File\ size\ after\ compression}{Original\ Fize\ Size} \tag{2}$$

Compression Time: Compression time indicates how much time was taken to compress a file. In this work, it was calculated in milliseconds (ms).

5.2   Result Comparison:

The outcome of applying the Run-length algorithm on the content of the files from Canterbury Corpus is given in Table 1.

Table 1. Comparison of Run Length Encoding on the contents of different text files from Canterbury Corpus

SN	File Name	Size (bytes)	Comp Size (bytes)	Comp Ratio	Saving Percentage	Comp Time (ms)
1	alice29.txt	1187848	4495984	0.2642	−278.5	11710.58
2	asyoulik.txt	1001432	3889168	0.2575	−288.36	8947.03
3	lcet10.txt	3353880	12608128	0.266	−275.93	92463.32
4	plrabn12.txt	3769296	14771744	0.2552	−291.9	126965.59

The outcome of applying LZW dictionary based compression algorithm on the content of the files from Canterbury Corpus is given in the Table 2 and the outcome of applying the Huffman compression algorithm on the content of the files from Canterbury Corpus is given in the Table 3.

Table 2. Comparison of LZW on the contents of different text files from Canterbury Corpus

SN	File Name	Size (bytes)	Comp Size (bytes)	Comp Ratio	Saving Percentage	Comp Time (ms)
1	alice29.txt	1187848	277896	4.2744	76.61	47.92
2	asyoulik.txt	1001432	250992	3.9899	74.94	42.24
3	lcet10.txt	3353880	669360	5.0106	80.04	94.84
4	plrabn12.txt	3769296	804176	4.6872	78.67	112.82

Table 3. Comparison of Huffman on the contents of different text files from Canterbury Corpus

SN	File Name	Size (bytes)	Comp Size (bytes)	Comp Ratio	Saving Percentage	Comp Time (ms)
1	alice29.txt	1187848	676374	1.7562	43.06	34.85
2	asyoulik.txt	1001432	606448	1.6513	39.44	32.28
3	lcet10.txt	3353880	1951007	1.7191	41.83	53.61
4	plrabn12.txt	3769296	2129465	1.7701	43.5	53.87

5.3 Result Analysis

From the previous section, it is apparent that among the three compression algorithms employed on the data set, LZW performed best. The performance was evaluated on the basis of compression ratio, space-saving percentage, and output file size. Although LZW performed better based on the aforementioned metrics, it took a little bit more time compared to Huffman when compressing files.

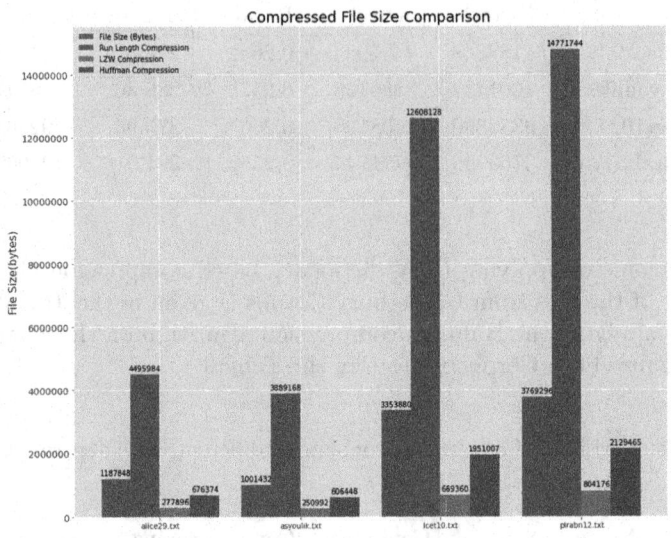

Fig. 1. Comparison of Compressed file size

On the contrary, RLE performed the worst all across the board due to its simplistic compression mechanism. The following figures compare the compressed file size, compression ratio, space saving percentage, and compression time of each of the algorithms using graphical representations.

Figure 1, shows the comparison of file size after applying each compression algorithm on the files of the aforementioned data set. In this graph, we can clearly see that RLE not only performed worse compared to the other two algorithms it performed significantly worse as we can see that the size of the compressed output was significantly bigger than the original file size. On the other hand, LZW performed best and gave the lowest output file size.

Figure 2, shows the comparison of compression ratio after applying each compression algorithm on the files of the aforementioned data set. In this graph, it can be seen that RLE also did not perform well compared to the other two algorithms. It performed significantly worse as we can see that the compression ratio values for RLE range from 0.2552 to 0.266 where the compression ratio value for LZW goes as high as 5.0106. Consequently, we can see that the compression ratio value of LZW is significantly higher than RLE. Huffman also gave a higher compression ratio ranging from 1.6513 to 1.7701 compared to RLE but it is still significantly lesser than LZW. Therefore LZW outperformed the rest of the algorithms when considering compression ratio as a metric.

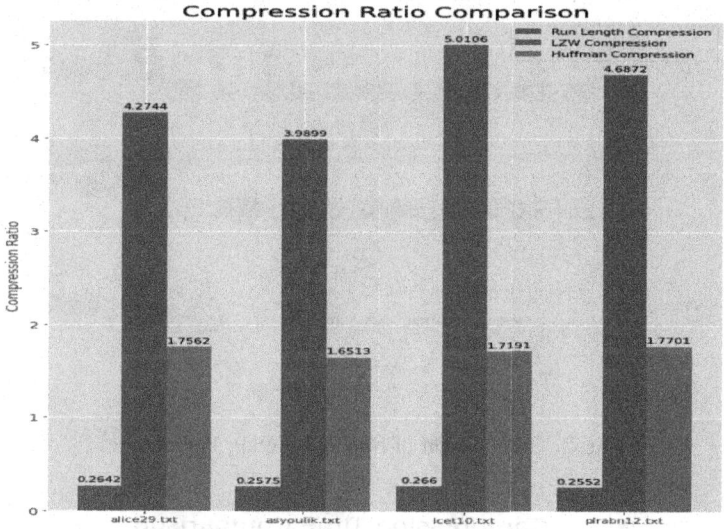

Fig. 2. Comparison of Compression Ratio

In Fig. 3, we can see the space saving percentage of each of the algorithms on the different files of the Canterbury Corpus. If we take a look at the results, RLE gave a very high negative value which indicates that the output file that we got after applying Run length encoding had a much bigger size compared to the input file so, it did not reduce any size rather increased the size and by a large factor. On the other hand, LZW gave space saving percentage values ranging from 74.94 to 80.04 and Huffman encoding provides space saving percentage values ranging from 39.44 to 43.06. Therefore, LZW also performed significantly better when considering space saving percentage as a metric.

According to Fig. 4, it can be seen that RLE took a large amount of time compared to the other two approaches. Compression time taken by RLE ranged from 8947.03 ms to as high as 126965.59 ms. Whereas LZW reaches 112.82 ms in case of the upper bound of its compression time. Finally, Huffman encoding provided 32.28 ms as the lowest value and 53.87 ms as the highest value. In case of the previous metrics, we spectated that LZW outperformed all the other compression algorithms apart from this case where Huffman performed better.

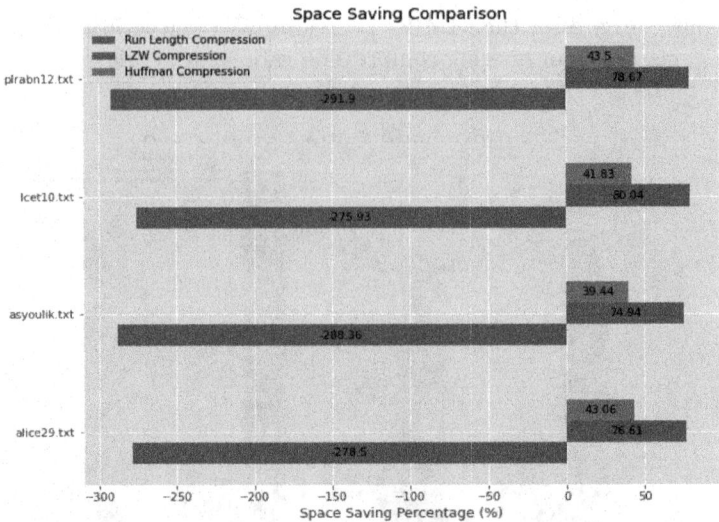

Fig. 3. Comparison of Spacing Saving Percentage

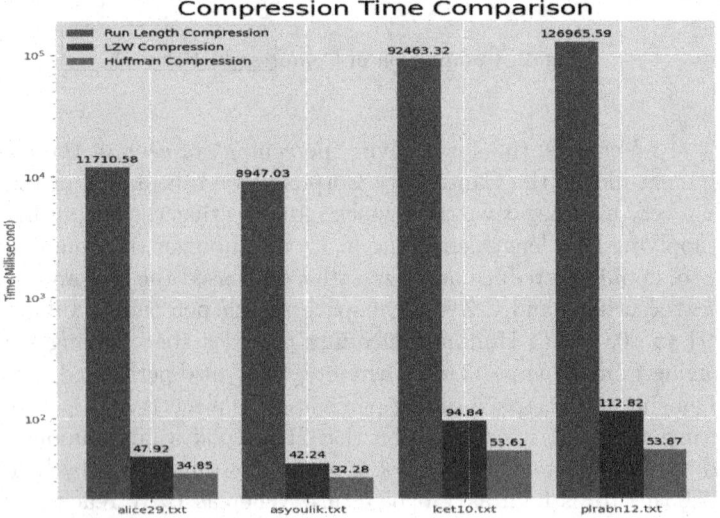

Fig. 4. Comparison of Compression Time

From the figures, it can be deduced that among the three compression algorithms LZW performed best in most categories except compression time, where Huffman performed slightly better than LZW. The tabular values also tell the same story. In each of the tables, we have calculated the compression ratio and space saving percentage for each of the input text files which are very important metrics to measure the effectiveness of compression algorithms. These metrics were calculated using 1 and 3 equations respectively.

6 Conclusion

In our experiment, our findings are as follows: RLE performed worse in every metric compared to the other two algorithms due to its simplistic compression mechanic as RLE can function effectively when there are repeating characters present which is not always applicable in real life scenarios. On the contrary, LZW performed better than others in most of the metrics except compression time due to its efficient technique of using dictionary. LZW could not outperform Huffman in one aspect which is compression time because LZW spends a substantial amount of time building the dictionary. In order to improve performance and efficiency, we intend to use an extended dataset and employ the ensemble modeling technique in the future.

References

1. Dobre, C., Xhafa, F.: Intelligent services for big data science. Futur. Gener. Comput. Syst. **37**, 267–281 (2014)
2. Sivarajah, U., Kamal, M.M., Irani, Z., Weerakkody, V.: Critical analysis of big data challenges and analytical methods. J. Bus. Res. **70**, 263–286 (2017)
3. Kodituwakku, S.R., Amarasinghe, U.S.: Comparison of lossless data compression algorithms for text data. Indian J. Comput. Sci. Eng. **1**(4), 416–425 (2010)
4. Porwal, S., Chaudhary, Y., Joshi, J., Jain, M., et al.: Data compression methodologies for lossless data and comparison between algorithms. Int. J. Eng. Sci. Innov. Technol. (IJESIT) **2**, 142–147 (2013)
5. Arnold, R., Bell, T.: A corpus for the evaluation of lossless compression algorithms. In: Proceedings DCC '97. Data Compression Conference, pp. 201–210 (1997)
6. Brisaboa, N.R., Cánovas, R., Claude, F., Martínez-Prieto, M.A., Navarro, G.: Compressed string dictionaries. In: Pardalos, P.M., Rebennack, S. (eds.) SEA 2011. LNCS, vol. 6630, pp. 136–147. Springer, Heidelberg (2011). https://doi.org/10.1007/978-3-642-20662-7_12
7. Brisaboa, N.R., Cerdeira-Pena, A., de Bernardo, G., Fariña, A., Navarro, G.: Space/time-efficient RDF stores based on circular suffix sorting. J. Supercomput. 1–41 (2022)
8. Kempa, D., Prezza, N.: At the roots of dictionary compression: string attractors. In: Proceedings of the 50th Annual ACM SIGACT Symposium on Theory of Computing, pp. 827–840 (2018)
9. Nishimoto, T., Tabei, Y.: LZRR: LZ77 parsing with right reference. In: 2019 Data Compression Conference (DCC), pp. 211–220 (2019)
10. Platos, J., Dvorský, J.: Word-based text compression. CoRR, abs/0804.3680 (2008)
11. Javed, M., Nagabhushan, P.: Automatic page segmentation without decompressing the run-length compressed text documents. arXiv preprint arXiv:2007.01142 (2020)
12. Fiergolla, S., Wolf, P.: Improving run length encoding by preprocessing. In: 2021 Data Compression Conference (DCC), pp. 341–341 (2021)

13. Shanmugasundaram, S., Lourdusamy, R.: A comparative study of text compression algorithms. Int. J. Wisdom Based Comput. **1**(3), 68–76 (2011)
14. Welch, T.A.: A technique for high-performance data compression. Computer **17**(06), 8–19 (1984)
15. Huffman, D.A.: A method for the construction of minimum-redundancy codes. Proc. IRE **40**(9), 1098–1101 (1952)

An Analysis of the Performance Changes of the Model by Reducing the Input Feature Dimension in Stock Price Forecasting

Yoojeong Song, Woo Jin Cho, and Juhan Yoo[✉]

School of Computer Science, Semyung University, 65, Semyeong-ro, Jecheon-si,
Republic of Korea
{yjsong,unchinto}@semyung.ac.kr

Abstract. In general, it can suffer from the course of dimensionality in developing machine learning models using high-dimensional data. To solve this problem, various dimensionality reduction algorithms have been developed, and PCA is the most widely used dimensionality reduction algorithm. In this paper, the PCA algorithm is applied to deep learning-based stock price prediction models developed in previous studies and their performance changes are analyzed. The models used data expected to affect stock price fluctuations as input features, and the data consist of 715 and 250 high-dimensional features, respectively. As a result, they not only took a lot of time in the model training and testing but also had the disadvantage of generating noises. Therefore, in this paper, the number of input features used in previous studies was reduced by using the dimension reduction method, and the performance change was analyzed. As a result of the experiment, it was confirmed that the models to which the PCA is applied have improved training speed and performance compared to the model without the PCA.

Keywords: Deep Learning · Stock Price Forecasting · Dimensionality Reduction · PCA Algorithm

1 Introduction

In general, it can suffer from the course of dimensionality in developing machine learning models using high-dimensional data. To solve this problem, various dimensionality reduction algorithms have been developed, and PCA is the most widely used dimensionality reduction algorithm [1]. In this paper, the PCA algorithm is applied to deep learning-based stock price prediction models developed in previous studies and their performance changes are analyzed. The models used data expected to affect stock price fluctuations as input features, and the data consist of 715 and 250 high-dimensional features, respectively. As a result, they not only took a lot of time in the model training and testing but also had the disadvantage of generating noises. Therefore, in this paper, the number of input features used in previous studies was reduced by using the dimension reduction method, and the performance change was analyzed. As a result of the experiment, it was confirmed that the models to which the PCA is applied have

F. Neri et al. (Eds.): CCCE 2023, CCIS 1823, pp. 171–178, 2023.
https://doi.org/10.1007/978-3-031-35299-7_14

improved training speed and perform. Through the COVID-19 pandemic in 2020, people have been interested in non-face-to-face money management. In this trend, stock
investment has become a representative money management means to increase their
property. According to a report by CNBC, 15% of about 500 investors started investing
in 2020, and JMP estimated that more than 7.8 million small new investors entered the
market in January and February 2021 due to the retail trading boom [2]. It is difficult for
most small investors to cope with these rapid market changes. Therefore, various stock
price prediction methods have been studied, and their goal is to develop a model that
can obtain a stable profitance compared to the model without the PCA.

For the stable stock price prediction, [3, 4] developed input features that can be used
in a deep learning-based prediction model by considering factors that may affect the
prediction of stock price fluctuations. The developed input features were constructed
based on technical indicators such as moving average, trading volume, and slope of
moving average and trading volume, including basic price data of stock. By using these
input features, the stock price prediction model was able to obtain a profit exceeding the
domestic price index and showed their novelty through various prediction experiments.
However, since the dimensionality of these input features is composed of 715 and 250,
respectively, models using these features not only take a lot of time to learn the data
but also have a problem in that noise occurs during training and testing. Therefore, in
this paper, the dimensionality of the input features used in the previous study is reduced
by using the PCA algorithm, and we compare and analyze the performance changes
between a model with reduced-dimensional features and a model with original features.

2   Related Works

2.1   Deep Learning Models for Stock Price Prediction

In general, methods using statistical models have been mainly used for stock price
prediction. However, recently, as the data prediction performance using deep learning
technology has improved, many stock price prediction methods based on this have been
studied. And there is a study that trading systems using reinforcement learning as well
as deep learning are effective in predicting stock prices [5]. According to this study
trend, [6] proposed three models using price-based input features. [3] introduced a 715-
dimensional model using advanced features based on technical analysis to improve the
prediction accuracy of stock price fluctuation patterns, and [4] proposed a model using
250-dimensional event binary features.

2.2   Dimensionality Reduction & PCA Algorithms with Stock

Input features are very important for developing machine learning models. However, it
can suffer from the curse of dimensionality [1] when the dimension of the input features
increases. The curse of dimensionality is a problem in which the number of training
data becomes smaller than the number of data dimensions, resulting in degradation of
performance. To solve this problem, a dimensionality reduction algorithm can be used,
and there are two representative dimensionality reduction methods. One is to keep only

the most relevant variables in the original data set (feature selection), and the other is to find a new set of variables smaller than the existing data set by using the redundancy of the input data. Among various methods for implementing this, a dimensionality reduction method using PCA is widely used. PCA is an algorithm that reduces the dimension of data in a high-dimensional space while maximally maintaining the variance between data. It can remove noise from data and eliminate unnecessary features [7]. In addition, PCA can be applied to deep learning-based models, and [8] improved the training efficiency by 36.8% by applying PCA to the LSTM model.

As mentioned above, [3, 4] proposed the stock price prediction model based on deep learning, but they can suffer from the curse of dimensionality by 715 and 250 high-dimensional input features, respectively. To solve this problem, in this paper, the dimensionality of the input features is reduced by using PCA, and we compare and analyze the stock price prediction performance between the model with reduced-dimensional features and the model with original features.

3 Model Design Using PCA Algorithms & Experiment

3.1 Detailed Design of the Model

The model used for the experiment consists of a deep neural network with two hidden layers. The number of nodes in the first and second hidden layers is determined experimentally, and as a result, they have 50 and 200 nodes, respectively. The final prediction target is set as a binary target that divides into two cases. One is a case where the price rises more than 10% after 5 days, and the other is a case where it does not. The stock price rise rate Rate is defined as follows:

$$\text{Rate} = \frac{High^s_{t+5}}{Close^s_t} \tag{1}$$

The prediction target is set to (0, 1) when is 10% or more, and is set to (1, 0) otherwise. The model is trained 2000 times with 500 batch size. And if the training performance has not improved more than 50 times, the training stops automatically. For this, EarlyStopping module [9] in Keras is used. RMSProp [10] is used as the optimization method, the hyperbolic tangent (Tanh) function is used as the activation function of the hidden layer, and the Softmax function [11] was used in the output layer. Finally, dropout is used to prevent overfitting, and the ratio is set to 0.5. Cross entropy is used as a loss function to predict a multi-interval target.

In this experiment, the model is trained using RTX2080 in Linux Ubuntu 18.04. The data used in the experiment is data on all stocks of the KOSPI/KOSDAQ from November 4, 2016, to September 6, 2019, and the composition is as follows Table 1.

Table 1. Data configuration of training, verification and test.

Model	Period	Note(days)
Training	2016–11-04 ~ 2018–06-27	about 600
Validation	2018–06-28 ~ 2019-01-31	about 200
Test	2019–02-01 ~ 2020-09-06	about 200

4   Experiments

4.1   The Performance of Prediction Models

To apply PCA, an eigen vector and an eigen value of each feature are calculated, and a threshold value of 0.95 is used. As a result, as shown in Table 2, 715-dimensional features are reduced to 248 dimensions, and 250-dimensional features are reduced to 183 dimensions.

Table 2. Reduced feature using PCA algorithms.

Input Feature	715 Advanced model	250 Event Binary model
Scree plot	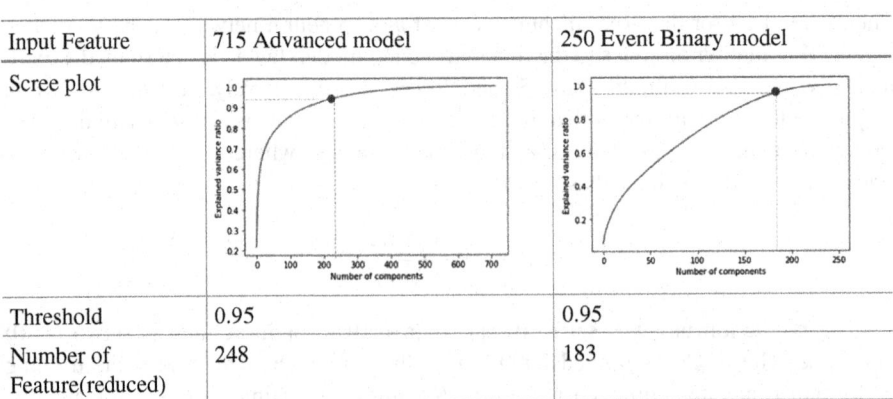	
Threshold	0.95	0.95
Number of Feature(reduced)	248	183

In this paper, the performances of four models are compared. Two of the models are original and the others are models with reduced-dimensional features by PCA. The training loss and the accuracy on test data are used as performance evaluation metrics for the models. The accuracy is a metric that can intuitively indicate a model's performance, but it cannot consider the bias of the data on the domain. Therefore, we use F1-Score [12] and AUC-Score [13] to compensate for this problem.

Table 3 shows the training loss and the accuracy of the 4 models. The accuracy of each model was more than 90%, and out of them, the reduced-dimensional 715 model (715-PCA model) had the highest accuracy and was about 7.59% higher than the original model (715 Advanced model). The accuracy of the model using the 250-dimensional event binary feature (250 Event Binary model) was 94.12%, and the accuracy of its reduced-dimensional model (250-PCA model) was 95.84%. Similar to the experimental results for the 715-dimensional feature model, the accuracy of the 250-PCA model was 1.72% higher than that of the 250 Event Binary model.

Table 3. Training results each models.

Model	Number of Feature	Training Loss	Accuracy(%)
715 Advanced model	715	0.2586	90.96
250 Event Binary model	250	0.1829	94.12
715 – PCA model	248	0.0700	98.55
250 – PCA model	183	0.1388	95.84

The experimental results for training loss and training accuracy are shown in Table 4, which shows that there is a difference in the training results of the models using PCA and the models without it. In the case of the models to which PCA was applied, the gradient of the training loss and the accuracy decreased and increased smoothly, but in the case of the models to which PCA was not applied, the training loss and the accuracy decreased and increased irregularly. Also, the models with reduced-dimensional features rapidly reduced the training loss even with a relatively small number of epochs. This means that noises of features are reduced and training speed is improved through PCA. However, in the case of the deep learning model, it is difficult to judge the performance of the model by using only the training accuracy. Therefore, the cross-validation is performed on test and validation data.

Table 5 shows that the accuracy of the trained model on the validation and the test data. On the verification data, the model using 715 advanced input features showed the highest accuracy, and the model using 250 event binary features showed the accuracy of 89.23%. However, the model with reduced dimensionality by PCA was less accurate than the original model.

On the test data, except for the model that reduced the dimensions of 250 event binary input features, all the other models showed high accuracy of over 99%. As a result, PCA can reduce the noise of the data and improve the training speed, however it does not affect the improvement of the prediction accuracy.

We further reevaluated the model using F1-Score and AUC-Score for more accurate performance evaluation. As a result, as shown in Table 6, it was confirmed that the model with all three indicators high was a model using 250 event binary input features. In addition, models using 715 advanced input features performed relatively lower than 250 event binary feature models, but performed better than two models using PCA.

Table 4. Training loss and accuracy each models.

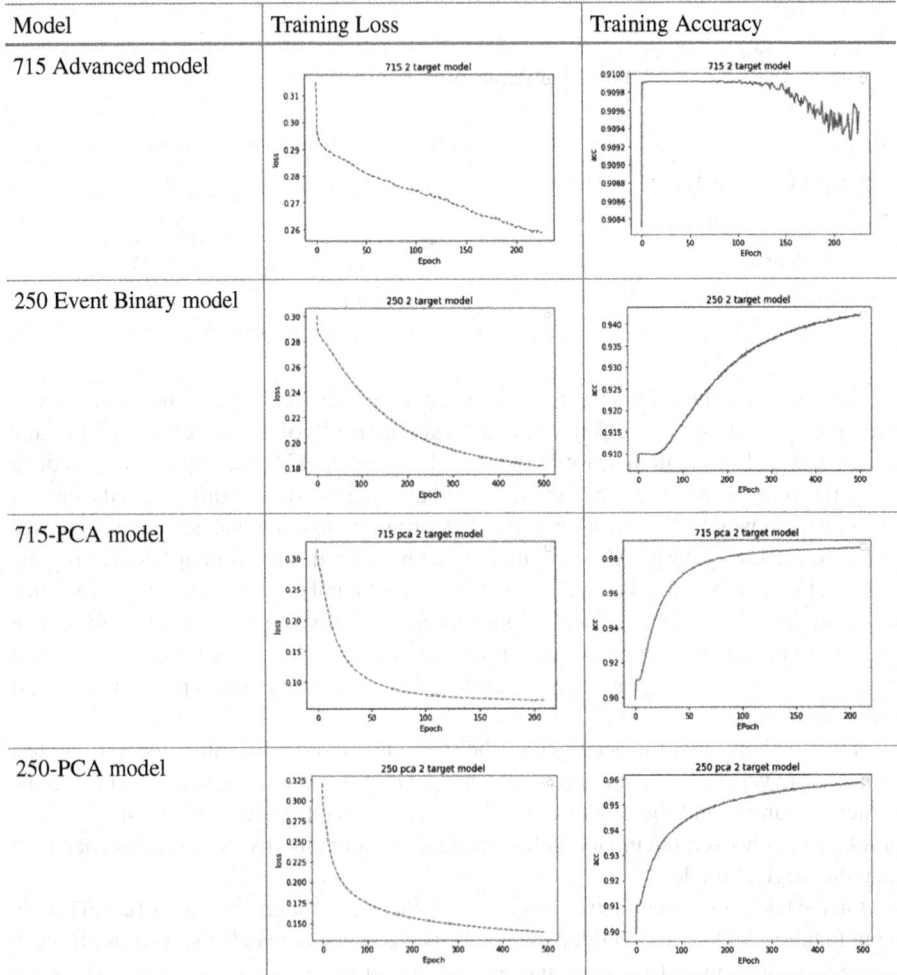

Model	Training Loss	Training Accuracy
715 Advanced model		
250 Event Binary model		
715-PCA model		
250-PCA model		

Table 5. Prediction error and accuracy of validation data and test data.

Model	Training Accuracy	Validation Accuracy	Test Accuracy
715 Advanced model	90.96	92.70	99.99
250 Event Binary model	94.12	89.23	99.99
715-PCA model	98.55	82.72	99.89
250-PCA model	95.84	85.05	81.06

(Unit of Training, Verification, Test accuracy is %)

Table 6. Model re-evaluation using F1-Score and AUC-Score.

Model	F1-Score	AUC-Score
715 Advanced model	0.8855	0.5726
250 Event Binary model	0.8896	0.6268
715–PCA model	0.8275	0.5208
250–PCA model	0.8645	0.5323

5   Conclusion

In this paper, PCA was applied to improve the performance of the models proposed by [3, 4]. First, 715 features and 250 features were reduced to 248 and 183 by using PCA, respectively. The training performance of the original models was compared with the models to which PCA was applied, and cross-validation was performed. Finally, the models were re-evaluated using the F-1 Score and AUC-Score. As a result of the experiment, the training accuracy of the 715-PCA model was 7.59% higher than that without applying PCA. And the training accuracy of the 250-PCA model was 1.72% higher than that without applying PCA. Therefore, the dimensionality reduction of input features by using PCA can help to reduce the data noise and improve training performance. However, the accuracies on the verification and the test data were low.

As a result, the dimensionality reduction for input features used in stock price prediction can help to improve training time and training performance, but there was no significant performance improvement in actual stock price prediction. This is because the prediction accuracy can be lowered by the multicollinearity problem [14] in which a strong correlation can appear between input features with reduced dimensions. In order to analyze this, the correlation analysis will be studied on training and test data as future work. Also, we will study a model that can improve the training performance as well as the stock price prediction accuracy by using not only PCA but also other dimensionality reduction techniques.

Acknowledgement. This research was supported by the Basic Science Research Program through the National Research Foundation of Korea (NRF) funded by the Ministry of Education (grant number: RS-2022–00165818).

References

1. Köppen, M.: The curse of dimensionality. In: 5th Online World Conference on Soft Computing In Industrial Applications (WSC5) (Vol. 1, pp. 4–8) (2000, September)
2. Maggie Fitzgerald, A large chunk of the retail investing crowd started during the pandemic, Schwab survey shows, CNBC, PUBLISHED THU, APR 8 2021 8:09 AM EDT UPDATED TUE, AUG 17 2021 11:39 PM EDT. https://www.cnbc.com/2021/04/08/a-large-chunk-of-the-retail-investing-crowd-got-their-start-during-the-pandemic-schwab-survey-shows.html

3. Song, Y., Lee, J.W., Lee, J.: A study on novel filtering and relationship between input-features and target-vectors in a deep learning model for stock price prediction. Appl. Intell. **49**(3), 897–911 (2018). https://doi.org/10.1007/s10489-018-1308-x
4. Song, Y., Lee, J.: Importance of event binary features in stock price prediction. Appl. Sci. (2020)
5. Rundo, F.: Deep LSTM with reinforcement learning layer for financial trend prediction in FX high frequency trading systems. Appl. Sci. (2019)
6. Song, Y., Won Lee, J., Lee, J.: Performance evaluation of price-based input features in stock price prediction using Tensorflow. KIISE Trans. Comput. Pract. **23** (11), 625–631 (2017)
7. Sorzano, C.O.S., Vargas, J., Montano, A.P.: A survey of dimensionality reduction techniques. arXiv preprint arXiv:1403.2877 (2014)
8. Shen, J., Omair Shafiq, M.: Short-term stock market price trend prediction using a comprehensive deep learning system. J. Big Data **7**, 66 (2020)
9. Bisong, E.: Regularization for deep learning. In: Building Machine Learning and Deep Learning Models on Google Cloud Platform (pp. 415–421). Apress, Berkeley, CA (2019)
10. Zou, F., Shen, L., Jie, Z., Zhang, W., Liu, W.: A sufficient condition for convergences of Adam and RMSProp. In: Proceedings of the IEEE/CVF Conference on computer vision and pattern recognition (pp. 11127–11135) (2019)
11. Chen, S., He, H.: Stock prediction using convolutional neural network. IOP Conf. Ser.: Mater. Sci. Eng. **435**(1), 012026. IOP Publishing (2018, October)
12. Chicco, D., Jurman, G.: The advantages of the Matthews correlation coefficient (MCC) over F1 score and accuracy in binary classification evaluation. BMC Genomics **21**(1), 1–13 (2020)
13. Lobo, J.M., Jiménez-Valverde, A., Real, R.: AUC: a misleading measure of the performance of predictive distribution models. Glob. Ecol. Biogeogr. **17**(2), 145–151 (2008)
14. Farrar, D.E., Glauber, R.R.: Multicollinearity in regression analysis: the problem revisited. The Review of Economic and Statistics, 92–107 (1967)

A Hybrid Algorithm by Incorporating Neural Network and Metaheuristic Algorithms for Function Approximation and Demand Prediction Estimation

Zhen-Yao Chen[✉]

Department of Business Administration, Hungkuo Delin University of Technology,
New Taipei City, Taiwan
keyzyc@gmail.com

Abstract. This study intends to heighten the training expression of radial basis function neural network (RbfN) via artificial neural network (ANN) and meta-heuristic (MH) algorithms. Further, the self-organizing map neural network (SOMN), artificial immune system (AIS) and ant colony optimization (ACO)-based algorithms are employed to train RbfN for function approximation and demand prediction estimation. The proposed hybrid of SOMN with AIS-based and ACO-based (HSIA) algorithm incorporates the complementarity of exploitation and exploration potentialities to attain problem settle. The experimental consequences have evidenced that AIS-based and ACO-based algorithms can be integrated cleverly and propose a hybrid algorithm which attempts for receiving the optimal training expression among corresponding algorithms in this study. Additionally, method appraisal consequences for two benchmark nonlinear test functions illustrate that the proposed HSIA algorithm surpasses corresponding algorithms in accuracy of function approximation issue and the industrial computer (IC) demand prediction exercise.

Keywords: Metaheuristic Algorithm · Self-organizing Map · Neural Network · Ant Colony Optimization · Artificial Immune System

1 Introduction

Artificial neural networks (ANNs) have intense imitating property while indigent in deduction and argumentation [1]. Next, owing to the simply constructed topology and the capability to evidence how learning carries on a strict manner, the radial basis function (Rbf) neural network (RbfN) has been broadly adopted as the general function approximator to resolve nonlinear problems [2]. Besides, the self-organizing mapping (SOM) [3] algorithm is one of the most universal ANN model based on the unsupervised rival learning pattern [4]. Since the SOM network (SOMN) can be seen as a clustering technique [5], and it is an unsupervised imitating ANN that employs neighborhood and topology to cluster correlative data into one category [6].

© The Author(s), under exclusive license to Springer Nature Switzerland AG 2023
F. Neri et al. (Eds.): CCCE 2023, CCIS 1823, pp. 179–188, 2023.
https://doi.org/10.1007/978-3-031-35299-7_15

Artificial immune system (AIS) contains several calculation intelligence procedures under the joint attribute of being based on principles and actions that normally happen at the degree of the immune system (IS). AISs are obtaining enhanced regard from scientific domain as a result of their potentiality to resolve compound optimization tasks in few realms [7]. Afterwards, a popular NI-based algorithm which has a random intrinsic metaheuristic (MH) algorithm (MHA) that has been implemented to conduct much compound optimization tasks is ant colony optimization (ACO) approach. The properties of this MHA involve the following: robustness, forceful response, decentralized calculation, and lightly incorporating with distinct algorithms, induces ACO to gain more successful optimal solutions [8]. Some MHAs sustain reduction of diversity consequences in little exploration, results all answers converge into certain local optimal responses. So they invalid the occasion of finding the global optimal regions [9].

Accordingly, the aim for this study is to adopt SOMN first with auto-clustering ability to decide the initial activation neuron parameters of the hidden layer on RbfN. Consequently, through taking the advantages of ANN model and MH algorithm, this research proposes the hybrid of SOMN with AIS-based and ACO-based (HSIA) algorithm to train RbfN. In the proposed HSIA algorithm, AIS-based approach actions exploitation in local district to refrain insufficient convergence (i.e., merely to acquire the second best feasible solution) and simultaneously to make the solution space more centralize. Also, ACO-based approach actions exploration in global district to refrain the early convergence, which would make the solution space more various. Lastly, the HSIA algorithm acquired the best inspection consequences in accuracy than other relevant algorithms in literature. Moreover, the practice of empirical industrial computer (IC) demand prediction practice evidences that the HSIA algorithm has higher preciseness than other corresponding algorithms and the Box-Jenkins models.

2   Literature Review

ANNs are the most effectual category of intelligent methods for nonlinear pattern recognition, which are the material driven and global district approximators [10]. Afterwards, the recursive least-squares (RLS) or gradient search algorithm is utilized to modulate the parameters of Gaussian functions and the joined weights to heighten the modelling potentiality. To gain an expected expression, a RbfN with a great size is constantly declared. But, such a great RbfN always claim a high challenge in estimation and constantly creates numerically unreliable. Although few methods have been recommended to alleviate this issue, the supplementing accomplished is not noticeable [11].

A SOMN is a nonlinear NN model [3]. Due to its rapid imitating, self-organized and topology inherence, literature evidences that the SOMN can be effectually utilized to reduce down initial layout options [12].

In addition, various assurance optimization algorithms sustain local optimal trap. The major attributes of these algorithms are dominant supposes absence and population dependency [13]. For example, the immune system (IS) can safeguard alive organisms from intruding bacteria or virus, while be suppling with a mutually proposition between lymphocytes and in certain B cells [14]. Then, Diao & Passino [15] in 2002 proposed a based on AIS algorithm for the RbfN structure and adaptation of parameters. In addition,

AIS adopts the superiority of studying principles in the airframe IS to form intelligence issue solvers [16]. Additionally, as a result of their scattered nature, AIS have been advanced concretely for resolving optimal dispersed and decision-making issues [17].

Subsequently, the ACO algorithm, introducing the pheromone concept that every ant rewards on its seek from preceding ants. Thus, for mining the global simplified path, pheromone consistence acts a necessary part for piloting the ants' conduct. In this process, efficacy and precision of algorithm are influenced via treating method of congregated information. Latter, the ACO algorithm has truly advanced potentiality of finding, excellent solution, as substantial as a stone steady, brief calculate implement. Therefore, many scholars' have presented methods which are facilitated for ACO, with intends to prevail the algorithm failing [8].

There are a various number of hybrid MHAs, which also may present very promising and excellence outcomes with correct solutions in many challenging learning topics [18]. In detail, the utility of global exploration incorporated with local exploitation not only has been demonstrated in software test data makeup but also in a variation of applications involving machine learning and its corresponding matters [19]. Moreover, the combined mode has been widely applied in many domains, such as finances [20], agribusiness [21], health [22], industry [23, 24], and etc. In the incorporated model, the superiorities of different forecast models are merged, and better forecast results are realized [25].

3   Methodology

The major point underlying SOMN is that RbfNs are local approximates, and the centres of local Rbf neurons are regulated to shift to the real centre in the meaning of feature expression [26]. Also, the conventional SOMN formulation involves a decaying neighborship width over time to generate a more greatly adjusted output mapping [27].

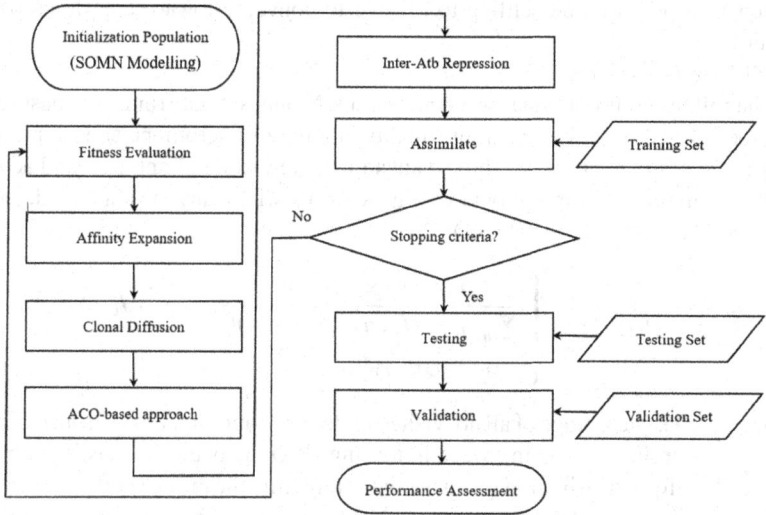

Fig. 1. The flowchart for the proposed HSIA algorithm.

Consequently, the flowchart of the HSIA algorithm is illustrated in Fig. 1, and a decla-
ration roughly on how the evolution program for the HSIA algorithm was explained as
follows. The proposed HSIA algorithm about how to train RbfN through ANN and MH
algorithms will be disserted in the inferior part.

(1) *Initialization Population*
(a) Initialize build a population which involve a random amount of antibody (Atb). Next,
    SOMN is adopted to judge the number of centres and the relevant position values by
    its auto-clustering talent. The outcomes are then applied as the number of neurons
    on RbfN. The Atbs emerged in the population can be regarded as the neurons during
    the hidden layer on RbfN. With two dissimilar benchmark nonlinear test functions,
    different input regions will bring unlike training patterns data (i.e., antigen (Atg))
    as well. These unlike training patterns data will be emerged within mapping input
    region on N dimension space for two benchmark nonlinear test functions.
(b) The weights w_i on RbfN are decided through resolving the linear system [28]:

$$Aw = u \tag{1}$$

where $A = A_{ij} = \xi_i(\|x - x_i\|_2)$ and $u = u(x_i)$ are the investigated function values at
the specimen points. The refined RBFs will consequence in a positive-definite matrix
$\mathfrak{R}$, accordingly insuring an individual solution to Eq. (1) [28]. Latter, the orthogonal
least squares (OLS) algorithm [29] is introduced to settle the weight value between
hidden and output layers on RbfN.

(2) *Fitness Evaluation & Affinity Expansion*
    The affinity expansion within Atb and Atg will be estimated during settling two
    benchmark nonlinear test functions. Afterward, an amount of candidate solutions
    with advanced fitness values will be acquired.

(3) *Clonal Diffusion*
    Such clonal diffusion action tolerates more Atbs to reproduce in the inherit popu-
    lation, and promote the settling to be able to converge toward the global optimal
    solution.

(4) *ACO-based approach*
(a) In the initiating of optimization sequence, let N ants set out from their nest to seek
    for food. It will initialize existing equality amount of pheromone at each margin of
    any ant's subsequent ways. Ants go ahead from any node of ant nest, and achieved
    at the terminal node throughout every generation, where anyone ant may decide the
    next node to visit through Eq. (2):

$$P_k(r, s) = \begin{cases} \dfrac{\tau(r, s)^\alpha \times \eta(r, s)^\beta}{\sum_{u \in J_k(r)} \tau(r, u)^\alpha \times \eta(r, u)^\beta}, & \text{if } s \in J_k(r) \\ 0, & \text{Otherwise} \end{cases} \tag{2}$$

where $J_k(r)$ is the storage of all not visited nodes once ant k continues from the initiate
node on margin r of passing ways in touring. $P_k(r, s)$ is the stochastic probability
value of uniform distribution for ant k continuing from the initiate node on margin r to
next stochastic visited node on margin s. Next, α and β are the experiential parameters
to manipulate the sensitiveness for pheromone value ($\alpha \geq 0$) and heuristic value

($\beta \leq 1$) respectively. $\tau_i(r, u)$ is the pheromone value from any node on margin r to any node on margin u, while $\eta(r, u)$ is the heuristic value from any node on margin r to any node on margin u.

(b) For the developing solution, it is inescapable that it will undergo with the local and global updates on pheromone examination.

(5) *Inter-Atb Repression*

During the evolution solving action, while the affinity within any two Atbs is lower than the repression threshold, one Atb will be restrained. The Atb who has excellent affinity with Atg will be more likely to determine the characteristics of data. Once it has proliferated Atbs with advanced affinity outspread within population, more memory elements (cells) will be attracted and will prompt secondary immune reaction to determine more Atgs.

(6) *Assimilate*

To preserve the equilibrium between steadiness and divergence within population in the process of resolving, Eq. (3) can be adopted to enlist a regular percentage (%) of Atbs in the progressive population:

$$N_{En} = \mu \cdot \tau \cdot Max[2^{-1} \cdot (F_1, Z_1), 50] \tag{3}$$

where (F_1, Z_1) is the quantity of sample points within training data in the experiment, $Max[2^{-1} \cdot (F_1, Z_1), 50]$ is the maximum quantity of Atbs within population. τ is the weakening element, and it may promote the steadiness in the process of solving. Further, μ is the enlist rate, and N_{En} is the enlistment quantity of Atbs within population. The amount of new attracted Atbs within the population is dictated through the assimilate action. By this process, it can refrain excessive attracted Atbs and make untimely divergence for the solution.

(7) *Update the global optimal solution*

While the stochastic emerged population has been investigated via the HSIA algorithm and data property has been correlated with the fitness function, the global optimal solution will be settled gradually.

(8) *Stopping criteria*

The proposed HSIA algorithm will keep on to realize and backtrack to *Step (2)* until the fitness function is fulfilled or a given number of generations is accomplished.

For the object of comprehend few unalike nonlinear optimization issues, this paper proposed the HSIA algorithm which incorporated the dominances of ANN and MH algorithms. The HSIA algorithm can enforce parallel processing with exploitation and exploration in the solution space, and prompt to refrain the solution from being stray the local optimal distress. In the principle, it allows the solution to converge by degrees and to gain the global optimal solutions. Eventually, the proposed HSIA algorithm can be applied on function approximation issue and the IC demand prediction exercise.

4   Experimental Evaluation Results

This section imploded on studying and modifying the pertinent parameters on RbfN for function approximation issue. The point is to gain the excellent advisable fitness values about the parameters solution of the RbfN. The emphasis is then to settle the apposite values of the parameters set from the searching area in the examination. The proposed HSIA algorithm will regulate and thus acquire the solution sets of parameters value for RbfN.

Experimental examination function prompts prominent approximation to reimburse RbfN for the reaction of nonlinear mapping relationship. This section employs two benchmark problems that are constantly adopted in the literature to be the comparative basis of estimated algorithms. Latter, the inspection involves the latter two benchmark problems, including Mackey-Glass time series [30] and Griewank [31] nonlinear test functions.

The first nonlinear test function, the Mackey-Glass time series [30] is expounded as follows:

$$\frac{dx(t)}{d(t)} = 0.1x(t) + 0.2 \cdot x(t - 17) \cdot [1 + x(t - 17)^{10}]^{-1} \tag{4}$$

where $x(t)$ is the value of time series at time step t [30]. In the other problem, Griewank nonlinear test function [31] is expounded as follows:

$$GR(x_j, x_{j+1}) = \sum_{j=1}^{n} \frac{x_j^2}{4000} - \prod_{j=1}^{n} \cos(\frac{x_{j+1}}{\sqrt{j+1}}) + 1 \tag{5}$$

where one global minimum is $(x_1, x_2) = (0, 0)$, and $GR(x_1, x_2) = 0$; the search region is $-100 \leq x_j \leq 100$ ($j = 1$).

4.1 Performance Estimation and Comparison

This paper combines ANN and MH algorithms to judge the optimal solution sets of parameters composition (the centre within neuron in hidden layer, width, and weight) for RbfN. According Looney [32] recommended in 1996, it stochastically produces unlike 65% training set from 1000 produces sample and distract the set into RbfN for training. By the alike action, it stochastically produces unlike 25% testing set to inquire into relevant parameters solution within population and assesses the fitness function. Until now, RbfN has introduced 90% data set in studying term. As a result of 1,000 iterations in the advance dealing, the optimal solution sets of parameters value for RbfN are gained. Lastly, it stochastically produces unlike 10% validation set to proof how the parameters solution of unit approximates the two appraisals and maintain the root mean square error (RMSE) values to clarify the studying term of RbfN. In case the data gathering term noticed above have realized, all algorithms are make ready to fulfil. The studying and validation terms referred above were carried out forty times before the average of RMSE values were evaluated. The evaluated consequences of the average of RMSE and standard deviation (SD) for all algorithms estimated from the investigation are announced in Table 1.

In Table 1, the announced outcomes evidence that HSIA algorithm acquires the reliable enough values with affirmative representation during the studying conduct of the exam checking. As the performance, RbfN may achieve the few solutions of parameters value set from the progressive procedure within population, which has carried out the circumstance with remarkable function approximation. When the modifying of RbfN via the HSIA algorithm is implemented, the unit with the optimal solutions of parameters value set during studying term is the RbfN situation in settled.

Table 1. Result examination among relevant algorithms applied in this experiment

Contrast Algorithms	Benchmark Functions			
	Mackey-Glass time series [30]		*Griewank* [31]	
	Training set	Validation set	Training set	Validation set
AIS-based [33]	2.88E-3 ± 1.68E-4	2.71E-3 ± 1.55E-4	15.63 ± 15.83	51.73 ± 16.65
ACO-based [34]	3.21E-3 ± 1.78E-4	2.67E-3 ± 2.08E-4	7.14 ± 10.30	49.14 ± 12.66
AIS-ACO [35]	2.26E-3 ± 1.91E-4	2.12E-3 ± 1.62E-4	6.59 ± 3.42	17.51 ± 4.37
HSIA	**1.85E-3 ± 1.58E-4**	**1.58E-3 ± 1.38E-4**	**4.77 ± 2.15**	**12.08 ± 2.19**

5   Case Study for Demand Prediction Estimation

This estimation tries to discuss the prediction precision on the sales specimen of the ROBO-series goods from 2008 to 2009 (227 tuples) offered by an internationally renowned industrial computer (IC) manufacturer in Taiwan. The SOMN and MH algorithms were applied to the proposed HSIA algorithm for sales specimen prediction confirmation inspection, and their precision was contrasted to relevant algorithms in literatures.

This case study expounds how specimen are feed in to RbfN for forecasting through corresponding algorithms and the contrast of results with the auto-regressive integrated moving average (ARIMA) [36] models. The specimen gathered in this practice are weekly sales specimen and thus are essential to be standardized to the training set. Consequently, $(n-4)^{th}$ to $(n-1)^{th}$ sales specimen are applied to forecast the nth tuple sales. After that, the following forecasted values are produced in turn by the moving-window method. The approximation expression of the RbfN forecast is inspected with the validation set (10% data). Besides, the first 90% of the specimen were adopted for model evaluation while the ultimate 10% were adopted for validation and one-step-ahead prediction. By the way, the RMSE, mean absolute error (MAE), and mean absolute percentage error (MAPE) are the majority frequently utilized error estimates in practice [37], and so were adopted to assess the prediction exercise. Thus, the prediction expressions among corresponding algorithms with the sales specimen of the ROBO-series goods are shown in Table 2.

The statistical consequences evidence that the HSIA algorithm has the best representation in terms of prediction preciseness among corresponding algorithms.

Table 2. The comparison of errors result among relevant algorithms applied in the IC case.

Contrast Algorithms	Errors		
	RMSE	MAE	MAPE(%)
AIS-based [33]	813.282	619.104	11.02
ACO-based [34]	637.923	548.257	9.71
AIS-ACO [35]	511.205	431.501	7.02
ARIMA models [36]	534.083	446.372	7.14
HSIA	**426.392**	**355.818**	**5.22**

6   Conclusions

This paper for the proposed HSIA algorithm via incorporating the SOMN, AIS and ACO-based approaches, which offers the enactment of RbfN parameters solution set, such as centre, width, and weight. The complementarity of ANN and MH algorithms that heightens the variety of populations also raises the precision of function approximation. The experimental estimation consequences have been contrasted with those achieved through the HSIA algorithm trained by several related algorithms. The HSIA algorithm has superior parameter setting and then causes RbfN to perform better learning and approximation in two benchmark nonlinear test functions. Additionally, an exercise case study has made an attempt to examine the prediction consequences on the sales specimen of the ROBO-series goods offered by an internationally renowned IC manufacturer in Taiwan. With the prediction preciseness confirmed, the verified HSIA algorithm can be applied to make forecasts in the realistic IC sales demand prediction exercise and could be gained for commerce decision managers to further heighten their benefit.

References

1. Shihabudheen, K.V., Pillai, G.N.: Recent advances in neuro-fuzzy system: a survey. Knowl.-Based Syst. **152**, 136–162 (2018)
2. Lin, G.F., Wu, M.C.: An RBF network with a two-step learning algorithm for developing a reservoir inflow forecasting model. J. Hydrol. **405**, 439–450 (2011)
3. Kohonen, T.: Self-Organizing and Associative Memory, 2nd edn. Springer, Berlin (1987). https://doi.org/10.1007/978-3-642-88163-3
4. Yadav, V., Srinivasan, D.: A SOM-based hybrid linear-neural model for short-term load forecasting. Neurocomputing **74**, 2874–2885 (2011)
5. Barreto, G.A., Araujo, A.F.R.: Identification and control of dynamical systems using the self-organizing map. IEEE Trans. Neural Netw. **15**(5), 1244–1259 (2004)

6. Chan, C.C.H.: Intelligent spider for information retrieval to support mining-based price prediction for online auctioning. Expert Syst. Appl. **34**, 347–356 (2008)
7. Galvez, A., Iglesias, A., Avila, A., Otero, C., Arias, R., Manchado, C.: Elitist clonal selection algorithm for optimal choice of free knots in B-spline data fitting. Appl. Soft Comput. **26**, 90–106 (2015)
8. Ebadinezhad, S.: DEACO: adopting dynamic evaporation strategy to enhance ACO algorithm for the traveling salesman problem. Eng. Appl. Artif. Intell. **92** (2020)
9. Salehpoor, I.B., Molla-Alizadeh-Zavardehi, S.: A constrained portfolio selection model at considering risk-adjusted measure by using hybrid meta-heuristic algorithms. Appl. Soft Comput. **75**, 233–253 (2019)
10. Hajirahimi, Z., Khashei, M.: Hybrid structures in time series modeling and forecasting: a review. Eng. Appl. Artif. Intell. **86**, 83–106 (2019)
11. Su, S.F., Chuang, C.C., Tao, C.W., Jeng, J.T., Hsiao, C.C.: Radial basis function networks with linear interval regression weights for symbolic interval data. IEEE Trans. Syst., Man, Cybernet.-Part B: Cybernet. **42**(1), 69–80 (2012)
12. Yan, W., Chen, C.H., Huang, Y., Mi, W.: An integration of bidding-oriented product conceptualization and supply chain formation. Comput. Ind. **59**, 128–144 (2008)
13. Pelusi, D., Mascella, R., Tallini, L., Nayak, J., Naik, B., Deng, Y.: An improved moth-flame optimization algorithm with hybrid search phase. Knowl.-Based Syst. **191** (2020)
14. Zhang, Z.: Fast multiobjective immune optimization approach solving multiobjective interval number programming. Swarm and Evol. Comput. **51** (2019)
15. Diao, Y., Passino, K.M.: Immunity-based hybrid learning methods for approximator structure and parameter adjustment engineering. Appl. Artif. Intell. **15**, 587–600 (2002)
16. Ozsen, S., Yucelbas, C.: On the evolution of ellipsoidal recognition regions in artificial immune systems. Appl. Soft Comput. **31**, 210–222 (2015)
17. Stogiannos, M., Alexandridis, A., Sarimveis, H.: An enhanced decentralized artificial immune-based strategy formulation algorithm for swarms of autonomous vehicles. Appl. Soft Comput. **89**, (2020)
18. Houssein, E.H., Saad, M.R., Hashim, F.A., Shaban, H., Hassaballah, M.: Levy flight distribution: a new metaheuristic algorithm for solving engineering optimization problems engineering. Appl. Artif. Intell. **94**, (2020)
19. Qian, C., Shi, J.C., Tang, K., Zhou, Z.H.: Constrained monotone K-submodular function maximization using multi-objective evolutionary algorithms with theoretical guarantee. IEEE Trans. Evol. Comput. **22**(4), 595–608 (2018)
20. Zhou, F., Zhou, H.M., Yang, Z.H., Yang, L.H.: EMD2FNN: a strategy combining empirical mode decomposition and factorization machine based neural network for stock market trend prediction. Expert Syst. Appl. **115**, 136–151 (2019)
21. Jimenez-Donaire, M.D., Tarquis, A., Giraldez, J.V.: Evaluation of a combined drought indicator and its potential for agricultural drought prediction in southern Spain. Nat. Hazard. **20**(1), 21–33 (2020)
22. Islam, M.J., Khan, A.M., Parves, M.R., Hossain, M.N., Halim, M.A.: Prediction of deleterious non-synonymous SNPs of human STK11 gene by combining algorithms, molecular docking, and molecular dynamics simulation. Sci. Rep. **9**, 16426 (2019)
23. Han, L., Huang, D., Yan, X., Chen, C., Zhang, X., Qi, M.: Combined high and low cycle fatigue life prediction model based on damage mechanics and its application in determining the aluminized location of turbine blade. Int. J. Fatigue **127**, 120–130 (2019). https://doi.org/10.1016/j.ijfatigue.2019.05.022
24. Wang, S.C., Liu, Z.T., Cordtz, R., Imran, M.G., Fu, Z.: Performance prediction of the combined cycle power plant with inlet air heating under part load conditions. Energy Convers. Manage. **200**, 112063 (2019)

25. Tian, Z.: Short-term wind speed prediction based on LMD and improved FA optimized combined kernel function LSSVM Engineering. Appl. Artif. Intell. **91**, 103573 (2020)
26. Er, M.J., Li, Z., Cai, H., Chen, Q.: Adaptive noise cancellation using enhanced dynamic fuzzy neural network. IEEE Trans. Fuzzy Syst. **13**(3), 331–342 (2005)
27. Rumbell, T., Denham, S.L., Wennekers, T.: A spiking self-organizing map combining STDP, oscillations, and continuous learning. IEEE Trans. Neural, Netw. Learn. Syst. **25**(5), 894–907 (2014)
28. Jakobsson, S., Andersson, B., Edelvik, F.: Rational radial basis function interpolation with applications to antenna design. J. Comput. Appl. Math. **233**(4), 889–904 (2009)
29. Chen, S., Cowan, C.F.N., Grant, P.M.: Orthogonal least squares learning algorithm for radial basis function networks. IEEE Trans. Neural Netw. **2**(2), 302–309 (1991)
30. Whitehead, B.A., Choate, T.D.: Cooperative-competitive genetic evolution of radial basis function centers and widths for time series prediction. IEEE Trans. Neural Netw. **7**(4), 869–880 (1996)
31. Bilal, M., Pant, H., Zaheer, L., Garcia-Hernandez, Abraham, A.: Differential evolution: a review of more than two decades of research. Eng. Appl. Artif. Intell. **90**, 103479 (2020)
32. Looney, C.G.: Advances in feedforward neural networks: demystifying knowledge acquiring black boxes. IEEE Trans. Knowl. Data Eng. **8**(2), 211–226 (1996)
33. Zhang, W., Yen, G.G., He, Z.: Constrained optimization via artificial immune system. IEEE Trans. Cybernet. **44**(2), 185–198 (2014)
34. Mavrovouniotis, M., Yang, S.: Ant colony optimization with immigrants schemes for the dynamic travelling salesman problem with traffic factors. Appl. Soft Comput. **13**(10), 4023–4037 (2013)
35. Savsani, P., Jhala, R.L., Savsani, V.: Effect of hybridizing biogeography-based optimization (BBO) technique with artificial immune algorithm (AIA) and ant colony optimization (ACO). Appl. Soft Comput. **21**, 542–553 (2014)
36. Box, G.E.P., Jenkins, G.: Time series analysis, forecasting and control. Holden-Day, San Francisco (1976)
37. Co, H.C., Boosarawongse, R.: Forecasting Thailand's rice export: statistical techniques vs. artificial neural networks. Comput. Ind. Eng. **53**, 610–627 (2007)

Deep QA: An Open-Domain Dataset of Deep Questions and Comprehensive Answers

Hariharasudan Savithri Anbarasu, Harshavardhan Veeranna Navalli[(✉)],
Harshita Vidapanakal, K. Manish Gowd, and Bhaskarjyoti Das

PES University, Bengaluru, Karnataka, India
harshavardhannavalli2@gmail.com

Abstract. Current available question answering (QA) datasets fall short in two aspects - providing comprehensive answers that span over a few sentences and questions being deep or analytical in nature. Though individually these issues are addressed, a dataset that addresses both these issues is still not available. To address this gap, we introduce Deep QA(DQA), i.e., a dataset consisting of 12816 questions broadly classified into 4 types of questions. The generated dataset has been analyzed and compared with a standard QA dataset to prove that it demands higher cognitive skills. To prove the point further, state of art models trained on remembering type factive QA dataset have been pre-trained on the proposed dataset and are shown to perform poorly on the question types generated. Finally, some preliminary investigation using a graph neural model has been done to probe the possibility of an alternative answer generation technique on such a dataset of deeper questions.

Keywords: Deep Questions · Comprehensive Answers · Semantic Roles · Proposition Bank · Neural QA · Graph based QA

1 Introduction

An important aspect of intelligence is the ability to understand and use qualitative knowledge. Understanding the semantic correlations and being able to use them in certain contexts are prerequisites for these skills. The development of algorithms that offer brief, concise replies has advanced significantly due to the availability of Question-Answering datasets. Open-ended questions that demand for such deeper understanding, however, have received less emphasis so far.

State of the art question answering models perform remarkably well on factual questions. There are abundant datasets available for factual/knowledge based Q/A such as SQuAD [1], XQuAD [2], SimpleQuestions [3] and EQG-Race [4]. There have also been cases where the questions are complex but provide brief responses such as KQA Pro [5], Mintaka [6] and GrailQA [7]. Subsequent work to improve the complexity of the question resulted in multi-hop reasoning based questions which was mainly possible due to datasets like HOTPOTQA [8]. Qualitative relationship or comparative questions have also been explored in depth in the NLP community. However, questions that require restatement of facts by relating the given knowledge to an entity and understanding the facts related to the entity require a higher level understanding.

To promote research based on this idea, we present the first open-domain dataset of comprehensive questions and answers, named DQA. The length and variety of responses that range multiple sentences is the key aspect of DQA. Complex questions that are not merely factual but are conceptual, procedural or meta cognitive, cannot simply be answered with a brief responses. These types of questions require a thorough comprehension of the context from which they were derived.

The rest of the article is organized as follows: Sect. 2 provides a comprehensive review of the evolution of question answering datasets highlighting the need for DQA. In Sect. 3, the DQA dataset, methods employed for the generation of DQA and its structure are described. In Sect. 4, the dataset is analyzed and compared to the other available datasets. The findings of how current state-of-the-art models perform on DQA is then presented in Sect. 5. The Sect. 6 concludes the discussion and indicates the next steps.

The DQA dataset has been made publicly[1] available for additional research.

2 Related Work

There have been abundant QA datasets that can be categorized based on evaluation ability, answer type, domain and question type [9].

Evaluation ability can be categorized into multi-hop, reasoning, summarization, conversational and reading comprehension. We argue that DQA overlaps with reasoning and reading comprehension due to the nature of the questions. Even though it can be argued that several reasoning datasets exist such as ARC [10], LogiQA [11], OpenBookQA [12], the answer type of these is Multi-Choice. BoolQ [13] is another reasoning dataset. It offers Yes/No questions. It requires a good amount of inference ability while the questions are factual in nature. Since the answers are limited to a single word answer, it does not require comprehensive and explanatory answers.

Many extractive question-answering datasets, including TREC [14], SQuAD [1], XQuAD [2], EQG-RACE [4], NewsQA [15], SimpleQuestions [3], SearchQA [16] and QuAC [17], limit the answer to a single word or phrase from the input and assess using an exact match with the ground truth span. The solution is typically part of a single paragraph, despite the fact that these datasets are not intentionally designed to be complicated. This kind of answer type is called cloze answer type. While the response is still a brief span, HotpotQA expands on this strategy by creating questions that force models to undertake multi-hop reasoning spanning several pages. The majority of understanding type questions are excluded since the response needs to be simple as it needs to be reproduced from the supporting information.

---

[1] https://doi.org/10.5281/zenodo.7538113.

MINTAKA [16] is a large, naturally-elicited, and multilingual question answering dataset which has complex questions. It includes inquiries that call for more than just a simple fact lookup, such as multi-hop, comparative, or set intersection queries but provides responses which are a word or a phrase and the answer does not cover a span of multiple sentences.

QuarTz [18] contains textual qualitative relationships. It evaluates a system's comprehension and application of textual qualitative knowledge in a novel environment. Similar to MINTAKA, it provides a single comparative-based answer which doesn't explain in detail.

MS MARCO [19] is a collection of crowdsourced answers to Bing questions. It has written responses that are only a single phrase long and have brief input portions. Larger multi-document online input from TriviaQA [20] was gathered using Bing and Wikipedia data. Most queries may be answered with a brief extractive period because the dataset is constructed from trivia. The fundamental problem with these datasets is that, although having at least one phrase replies, they don't offer complex questions.

It is possible to think of the ELI5 [21] which creates a paragraph-length answer from a variety of relevant evidence as a type of query-based multi-document summarizing. As opposed to writing on a broad subject, it calls for more targeted text generation. DUC 20042 and other summarization tasks need the production of several sentences from lengthy input. Composing Wikipedia entries is suggested as the multi-document summarizing effort by a similar dataset i.e., WikiSum.

Clearly a dataset offering deeper questions with answers providing comprehensive explanations is lacking. A survey paper on 'reading comprehension' datasets [9] points out a lack of 'Explain why' questions that we address in DQA.

3   Dataset and Dataset Generation

The dataset consists of 12816 questions classified as deep questions according to the Bloom's Taxonomy. The questions belong to various areas in the field of computer science. They were extracted from Wikipedia summaries and text-books related to the field of computer science. Each row in the dataset is a question-answer-context triple. The answers contain one or more sentences. The dataset is mainly split into 4 question types: *'Explain why'*, *'Justify with an example'*, *'Describe'* and *'What is'*. It also contains other types of questions such as *'Describe the role of'* and *'List the used of'*. A sample question from our dataset is shown in Fig. 1.

Question Type	Question	Answer	Context
Explain why	Explain why calculations that are exact in decimal can be close but inexact in binary, which can be a problem for financial	One reason to use decimal operands is to get results that exactly match decimal numbers, as some decimal fractions do not have an exact representation in binary. For example, 0.1010 is a simple fraction in decimal but in binary it requires an infinite set of repeating digits: 0.0001100110011...2. Hence, calculations that are exact in decimal can be close but inexact in binary, which can be a problem for financial transactions.	For business applications, some architectures support a decimal format, usually called packed decimal or binary-coded decimal4 bits are used to encode the values 09, and 2 decimal digits are packed into each byte. Numeric character strings are sometimes called unpacked decimal, and operationscalled packing and unpackingare usually provided for converting back and forth between them. ; Type and Size of Operands One reason to use decimal operands is to get results that exactly match decimal numbers, as some decimal fractions do not have an exact representation in binary. For example, 0.1010 is a simple fraction in decimal but in binary it requires an infinite set of repeating digits: 0.0001100110011...2. Thus, calculations that are exact in decimal can be close but inexact in binary, which can be a problem for financial transactions. (See Appendix G to learn more about precise arithmetic.) Our SPEC benchmarks use byte or character, half word (short integer), word (integer), double word (long integer) and floating-point data types. Figure 2.12 shows the dynamic distribution of the sizes of objects referenced from memory for these programs. The frequency of access to different data types helps in deciding what types are most important to support efficiently. Should the computer have a 64-bit access path, or would taking two cycles to access a double word be satisfactory? As we saw earlier, byte accesses require an alignment network: How important is it to support bytes as primitives?

Fig. 1. Snippet of our dataset.

3.1 Domain

The dataset has been created using information from Wikipedia and contexts from numerous textbooks on computer science and science. The questions were generated from the PDF version of the textbooks available on online e-book libraries. The produced questions were then thoroughly screened to remove any meaningless or irrational ones and ensure that every question could be answered.

3.2 Questions Requiring Deeper Understanding

The knowledge dimension matrix by Cannon and Feinstein [22], as shown in Table 1, indicates the different complexity ranges of questions based on question-words or key words in the question. The columns represent the cognitive process dimension of the question that follows the classification specified by Bloom's Taxonomy whereas the rows represent the knowledge dimensions. The intersection of factual and remembering can be considered as Level 0 questions. The rest are considered 'deep questions' requiring deeper understanding.

Table 1. The knowledge dimension matrix filled with key indicators [26].

The Knowledge Dimension	The Cognitive Process Dimension				
	Level 1 Remember	Level 2 Understand	Level 3 Apply	Level 4 Analyze	Level 5 Evaluate
A Factual	name, list define, label	restate order	state determine	distinguish classify	select according to
B Conceptual	identify locate	describe explain	illustrate show	examine analyze	rank compare
C Procedural	tell describe	summarize translate	solve demonstrate	deduct diagram	conclude choose
D Meta-Cognitive	–	interpret paraphrase	find out use	infer examine	justify judge

In DQA, two of the four question types generated belong to the conceptual-understanding category. The other two belong to meta cognitive-evaluation category. We thus argue that DQA is the first dataset to consist of 'deep questions with comprehensive answers' that are not of cloze type.

3.3 Method for Generation of Questions

Different techniques have been used to create the different types of questions present in DQA:

– **Explain type and Justify type** questions follow a fairly straightforward method of generation. These questions are generated from pdf version of textbooks related to computer science. The pdf is converted to text using the calibre e-book converter. The converted text is then used to generate the questions. The text is examined and it is initially determined if forming such a question is possible based on words occurring in the context. If it is, the answer and question is formed based on extracting a relevant span of the context for each separately.

Figure 2 explains the generation in detail. A question sentence is any sentence which starts which a certain prefix and can be converted into an actual question by replacing the prefix. In the case of 'Explain why' type questions, we search for sentences having starting with either 'Therefore,', 'Thus,' or 'So,' as the question sentence. We then replace the prefix with 'Explain why'. We then form the answer for the question using the text above the question sentence in the current paragraph and also the text above the current paragraph if the answer turns out to be too small. The context is formed by combining the current paragraph and paragraphs above and below the current one.

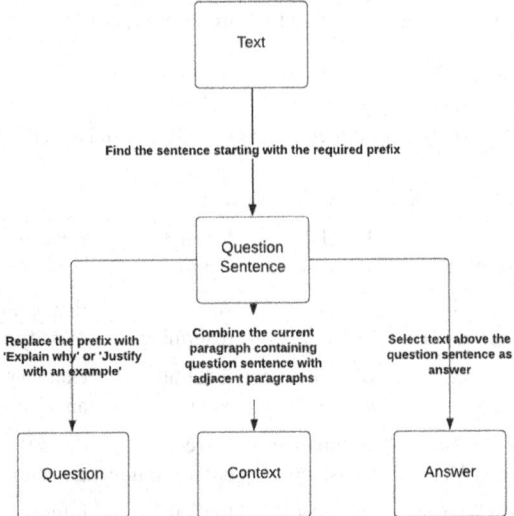

Fig. 2. Generating template-based questions.

Some of the questions that were generated had pronouns which could not be resolved using the question itself. AllenNLP Neural Co-reference model has been used for the resolution. For example. Consider the question i.e., 'Explain why it optimizes CPU utilization'. After resolving the co-reference using the context and the answer, we obtain the improved question as 'Explain why the working-set model optimizes CPU utilization'.

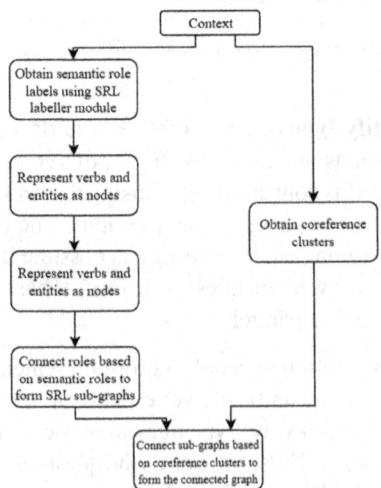

Fig. 3. Forming a connected a graph.

- **The other question types** are generated from Wikipedia summaries related to computer science using Semantic Role Labeling (SRL) [23]. The primary verb of a phrase is referred to as the predicate in linguistics. Predicate accepts justifications. To ascertain how these arguments are semantically associated to the predicate, **Semantic Role labelling (SRL)** is used. The **Proposition Bank** [24], often known as PropBank, is a way to carry out general purpose semantic role labelling with the help of a large annotated corpus where a set of semantic roles is tagged for each verb. In PropBank, the arguments of the verbs are labeled as numbered arguments: Arg0, Arg1, Arg2 and so on. Using the knowledge of Propbank, we were able to generate the following types of questions using the following heuristics:

  - **Describe:** The number of stative verbs must be greater than or to equal half of the total number of verbs in the spanned answer.
  - **Describe the role of:** The number of verbs under the category ARG0 must be greater than 70% of the total verbs and the number of verbs which are not in past tense and under the category ARG0 must be greater than 55% of the total number of verbs.
  - **Describe what happens to:** The number of verbs under the category ARG1 must be greater than 3 and the number of verbs under category ARG1 greater than 75% of total number of verbs.
  - **What is:** It is similar to "Describe", but with a maximum answer length of three lines.

- **List the uses of:** Observance of the occurrence of many verbs associated with the same entity. The verb with respect to which the entity is associated with should be of one of the following types: [ARG0, ARG1, ARG2, ARG3 etc.].

  By mentioning stative verbs, we refer to verbs that describe the state of an entity. The above-mentioned flow to generate other types of questions is explained in Fig. 3 and Fig. 4.

4 Dataset Analysis

In this analysis we try to characterize the complexity of the questions by comparing it with the famous large-scale dataset – SquAD [1].

4.1 Entities

It is observed that the questions in our dataset contain almost twice the number of entities in the questions belonging to the SQuAD dataset. This indicates the questions are complex involving more entities. The number of entities in contexts of our dataset contain more than 5 times of that in the contexts of SQuAD and the answers contain 20 times the entities when our dataset is compared with SQuAD.

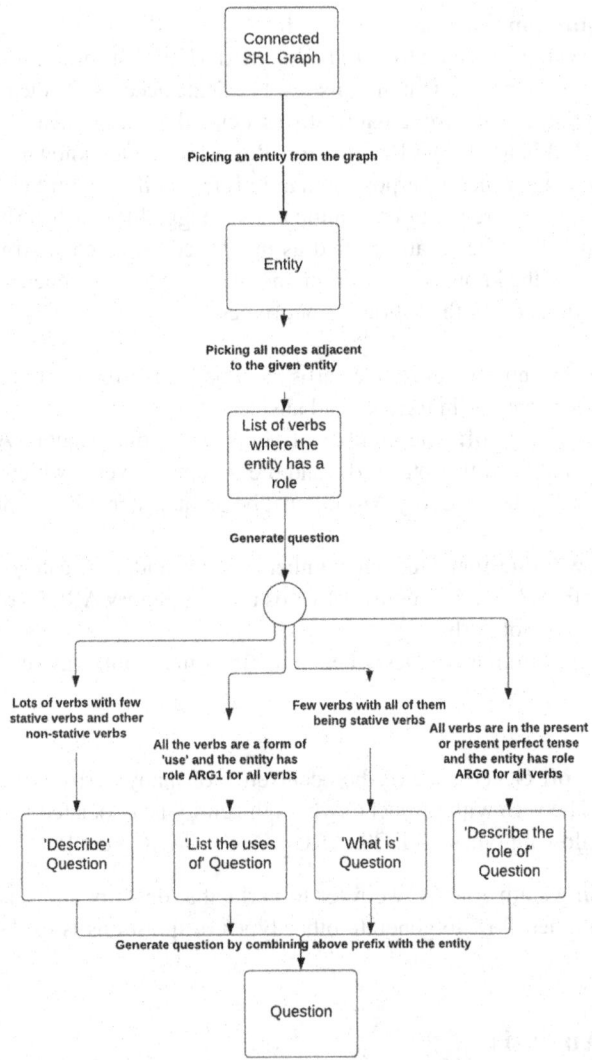

Fig. 4. Generation of question based on verbs.

4.2 Words

Here we can observe that the contexts are much larger compared to SQuAD. The questions are twice the size of those in SQuAD making the questions more descriptive or explanatory in nature. The average number of words in the answers is extremely large compared to that of SQuAD. This is due to the fact that the answers in SQuAD are mere phrases but answers in our dataset are multi-sentence answers.

Table 2. Comparative analysis with SQuAD.

Features	Our Dataset	SQuAD
Avg. entities/question	6.02	3.41
Avg. entities/context	228.84	42.97
Avg. entities/answer	33.46	1.65
Avg. words/question	22.80	11.40
Avg. words/context	794.55	140.27
Avg. words/answer	121.86	3.48
Avg. sentence/question	1.11	1.00
Avg. sentences/context	35.64	5.03
Avg. sentences/answer	5.17	1.00
Question avg. flesch reading ease score	50.56	70.15
Question avg. automatic readability index	12.42	7.32
Context avg. flesch reading ease score	51.36	46.51
Context avg. automatic readability index	12.52	15.62

4.3 Sentences

The answers of SQuAD containing mere phrases form a single sentence compared to the multi-sentence answers present in our dataset with an average of 5 sentences per answer. The sentences in the contexts of our dataset are 7 times higher than the number of sentences present in those of SQuAD dataset.

4.4 Readability Scores

Readability scores and indices determine how difficult it is to read and understand a given text. A high score or low index indicates that the text can be easily read and understood. The quality of the question in terms of the difficulty to read and understand can be determined using these metrics.

In Table 2, we measure the average readability scores of questions and contexts. We observe that the questions in our dataset achieve a much lower average Flesch score and much higher average automatic readability index compared to the questions in the SQuAD. This indicates that the questions in our dataset are more complex in terms of the readability making it harder to read and understand the question.

The average automatic readability index of the questions in our dataset is 12.42 meaning that a student should be at least in grade 12 to read and understand the question easily. Comparing that with the questions of SQuAD having a index of 7.32, it requires the student to be in at least grade 6 which is quite low compared to the questions in our dataset.

We observe that the contexts in our dataset have a higher flesch score with a difference of 4.85 and a lower readability index with a difference of 3.1 when compared to those of SQuAD. This makes the contexts in our dataset slightly easier to read and understand that the ones present in SQuAD.

Table 3 compares our dataset with SQuAD, based on the top 30 words from each of the datasets. The top 30 words are choosen based on the frequency of occurrence of the words where the word with the highest frequency would be at the top.

The words which are related to the field of computer science are in bold font and others are normal font. We can observe that our dataset contains words such as *data, system, process, network, information, table, model, database, algorithm, state* and *systems* which are related to computer science whereas SQuAD does not contain words related to the same field. SQuAD contains words like *city, world, century, united, people* and *government* which are not scientific and mostly used in historic or non-scientific facts. We can also observe that our dataset contains words such as *used, example, value, based* and *different* which are indicative of higher levels of cognitive thinking whereas SQuAD contains words such as *known* and *became* which are indicative of lower levels of cognitive thinking such as remembering facts.

From the above analysis, we can conclude that our dataset is **more scientific** and requires **higher levels of cognitive thinking** whereas SQuAD is non-scientific and factual.

Table 3. Comparison between top 30 most occurring words from our dataset and SQuAD. The words that are related to the field of computer science are highlighted in bold.

Top 30 Words					
Our Dataset			SQuAD		
data	one	time	also	one	first
system	used	number	new	city	many
process	figure	two	two	world	time
set	also	**network**	state	states	used
may	example	use	may	century	united
using	**information**	**table**	war	would	known
thus	based	**model**	people	including	years
database	value	first	government	however	became
algorithm	**state**	different	year	use	well
systems	therefore	new	early	later	since

5 Evaluating the Answerability of DQA Questions

We evaluate the complexity of the DQA dataset by running state of art QA models for factual remembering type of questions and compare the closeness of the generated answers to the actual answers in DQA. This reasoning is similar to the method used in the paper [25] where the Ontology for Vietnamese Language has been evaluated using a question answering system. We evaluate four baseline models on DQA.

- **BART:** BART model is remarkably suitable when fine-tuned for text generation, and it also performs admirably for comprehension tasks (e.g. question answering). The model was fine-tuned on SQuADv2 dataset [26].
- **BERT:** The BERT model was fine-tuned from the HuggingFace BERT base uncased on SQuAD1.1. This model is case-insensitive [27].

Table 4. The BERT metric is run on the actual answer and the answer predicted by the baseline models with the context to check the answer-ability of the models.

	Explain Why		Justify		Describe		What is	
	Actual	Predicted	Actual	Predicted	Actual	Predicted	Actual	Predicted
BERT	0.73	0.37	0.76	0.37	0.68	0.36	0.74	0.34
T5	0.73	0.35	0.76	0.34	0.68	0.58	0.74	0.61
BART	0.73	0.38	0.76	0.31	0.68	0.34	0.74	0.34

- **T5:** Google's T5-base fine-tuned on BioASQ (secondary task) for QA down- stream task [28].
- **BERT (Fine Tuned):** BERT model which was fine tuned on our own dataset.

Encoder-decoder models like T5 and BART are frequently employed in these situations to generate the information in a manner which is quite comparable to text summarization. Since some encoder-only models excel at extracting responses to factoids or knowledge-based questions but struggle when faced with complex questions requiring deeper understanding, BART and T5 were recognized as the best options.

On fine tuning our dataset on BERT Question Answering model, we found that it gives a BLEU score of 0.40 which is only a 0.04 increase in comparison to the model's performance without fine tuning as shown in Table 4.

The models were fine-tuned on popular QA databases and the ones with the best results were chosen. To assess how accurately the baseline models have predicted answers given the topic and context, we employed BERT metrics. It was observed that 'Justify' question type was predicted poorly and 'What is' was predicted comparatively better. There was 30% difference indicating that DQA answers are significantly better than the baseline models. The results of the same are shown in Table 4. Poor performance in the observations indicate that the existing factual QA models could not adequately respond to the question at hand.

5.1 A Graph Based Approach for Answering Deeper Questions

Considering the poor performance of transformer based models to provide comprehensive answers to deeper questions, some preliminary investigations have been done using a graph based approach with a Graph Attention Network (GAT) [29]. The basis of this model is a modification of AS2 (Answer Selection) task[30]. AS2 aims to pick the most appropriate answer from a group of potential answers from the context. The workflow of the GAT model has been shown in Fig. 5.

Context information is captured by encoding sentences using transformer-based models. Encoded sentences are connected based on several inter-sentence relationships to form the edge attribute vector:

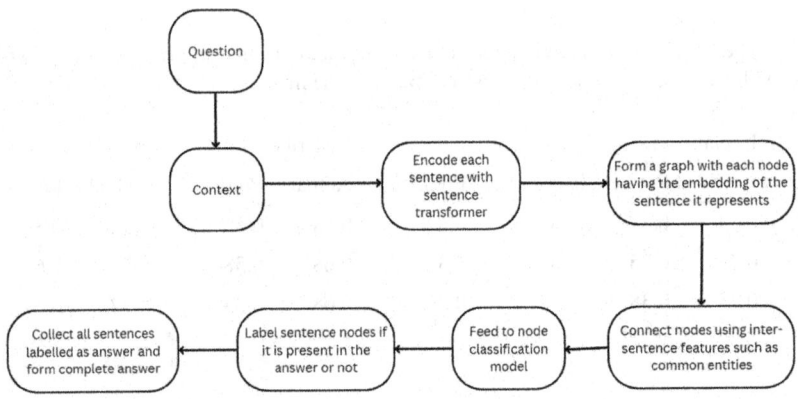

Fig. 5. GAT model workflow.

1. Common entities between pairs of sentences

$$score = \frac{Number\ of\ common\ entities\ between\ the\ two\ sentences}{Maximum\ number\ of\ entities\ among\ the\ two\ sentences}$$

Here entities refer to all the words in a sentence which are not stop words.

2. Inter-sentence distance - Absolute Value of the difference of the position of the pairs of sentences in the context
3. Adjacent sentence pointer - If both sentences connected by the edge are adjacent in the context
4. Question-sentence pointer - If the edge involves the question sentence
5. Cosine similarity between the embeddings of the sentences

The resultant graph is fed to the GAT node classification model and all the nodes/sentences labelled as a part of the answer are combined to form a complete answer. We use 3 GAT layers paired with 3 linear layers. The GAT layers capture the inter-sentence relation information effectively. The performance of this model is shown in Table 5. Its promising results set a precedent that an approach on similar lines with further improvements can yield better results.

Table 5. GAT model performance on DQA.

	ROC-AUC (Weighted)	Precision (Weighted)	Recall (Weighted)	MCC	F1 Score (Weighted)	Cohen Kappa
Our Model	74.97	89.47	78.78	31.62	77.43	24.20

6   Conclusion

In this paper, we provide DQA, the first open-domain dataset for question-answering that showcase deeper questions requiring large comprehensive answers that are so far not addressed by existing QA datasets. The analysis performed show that the DQA dataset require deeper cognitive ability as compared with the existing factual QA datasets. We also demonstrate that such deeper questions are difficult to answer by the state of art factual QA models. A graph neural based answering model has also been investigated to show that this is the direction adopt for such a dataset. Future plans include for the addition of other kinds of comprehension questions as well as improving the graph based answer generation models that it warrants.

References

1. Rajpurkar, P., Zhang, J., Lopyrev, K., Liang, P.: Squad: 100,000+ questions for machine comprehension of text. arXiv preprint arXiv:1606.05250 (2016)
2. Artetxe, M., Ruder, S., Yogatama, D.: On the cross-lingual transferability of mono-lingual representations. arXiv preprint arXiv:1910.11856 (2019)
3. Bordes, A., Usunier, N., Chopra, S., Weston, J.: Large-scale simple question answering with memory networks. arXiv preprint arXiv:1506.02075 (2015)
4. Jia, X., Zhou, W., Sun, X., Wu, Y.: Eqg-race: examination-type question generation. In: Proceedings of the AAAI Conference on Artificial Intelligence, vol. 35, pp. 13143–13151 (2021)
5. Cao, S., et al.: KQA pro: a dataset with explicit compositional programs for complex question answering over knowledge base. In: Proceedings of the 60th Annual Meeting of the Association for Computational Linguistics (ACL 2022), vol. 1:(Long Papers), pp. 6101–6119. Association Computational Linguistics-ACL (2022)
6. Sen, P., Aji, A.F., Saffari, A.: Mintaka: a complex, natural, and multilingual dataset for end-to-end question answering. arXiv preprint arXiv:2210.01613 (2022)
7. Gu, Y., et al.: Beyond IID: three levels of generalization for question answering on knowledge bases. In: Proceedings of the Web Conference 2021, pp. 3477–3488 (2021)
8. Yang, Z., et al.: Hotpotqa: a dataset for diverse, explainable multi-hop question answering. arXiv preprint arXiv:1809.09600 (2018)
9. Wang, Z.: Modern question answering datasets and benchmarks: a survey. arXiv preprint arXiv:2206.15030 (2022)
10. Boratko, M., et al.: A systematic classification of knowledge, reasoning, and context within the arc dataset. arXiv preprint arXiv:1806.00358 (2018)
11. Liu, J., Cui, L., Liu, H., Huang, D., Wang, Y., Zhang, Y.: LogiQA: a challenge dataset for machine reading comprehension with logical reasoning. arXiv preprint arXiv:2007.08124 (2020)

12. Mihaylov, T., Clark, P., Khot, T., Sabharwal, A.: Can a suit of armor conduct electricity? A new dataset for open book question answering. arXiv preprint arXiv:1809.02789 (2018)

13. Clark, C., Lee, K., Chang, M.W., Kwiatkowski, T., Collins, M., Toutanova, K.: Boolq: exploring the surprising difficulty of natural yes/no questions. arXiv preprint arXiv:1905.10044 (2019)

14. Voskarides, N., Li, D., Panteli, A., Ren, P.: ILPS at TREC 2019 conversational assistant track. In: TREC (2019)

15. Trischler, A., et al.: NewsQA: a machine comprehension dataset. arXiv preprint arXiv:1611.09830 (2016)

16. Dunn, M., Sagun, L., Higgins, M., Guney, V.U., Cirik, V., Cho, K.: SearchQA: a new Q&A dataset augmented with context from a search engine. arXiv preprint arXiv:1704.05179 (2017)

17. Choi, E., et al.: Quac: question answering in context. arXiv preprint arXiv:1808.07036 (2018)

18. Tafjord, O., Gardner, M., Lin, K., Clark, P.: Quartz: an open-domain dataset of qualitative relationship questions. arXiv preprint arXiv:1909.03553 (2019)

19. Nguyen, T., et al.: MS marco: a human generated machine reading comprehension dataset. Choice **2640**, 660 (2016)

20. Joshi, M., Choi, E., Weld, D.S., Zettlemoyer, L.: TriviaQA: a large scale distantly supervised challenge dataset for reading comprehension. arXiv preprint arXiv:1705.03551 (2017)

21. Fan, A., Jernite, Y., Perez, E., Grangier, D., Weston, J., Auli, M.: Eli5: long form question answering. arXiv preprint arXiv:1907.09190 (2019)

22. Ullrich, S., Geierhos, M.: Using bloom's taxonomy to classify question complexity. In: Proceedings of the Fourth International Conference on Natural Language and Speech Processing (ICNLSP 2021), pp. 285–289 (2021)

23. Palmer, M., Gildea, D., Xue, N.: Semantic role labeling. Synth. Lect. Hum. Lang. Technol. **3**(1), 1–103 (2010)

24. Palmer, M., Gildea, D., Kingsbury, P.: The proposition bank: an annotated corpus of semantic roles. Comput. Linguist. **31**(1), 71–106 (2005)

25. Nguyen, D., Nguyen, T.: A question answering model based evaluation for OVL (ontology for Vietnamese language). Int. J. Comput. Theory Eng. 347–351 (2011). https://doi.org/10.7763/IJCTE.2011.V3.330, https://huggingface.co/ChuVN/bart-base-finetuned-squad2

26. https://huggingface.co/ChuVN/bart-base-finetuned-squad2

27. https://huggingface.co/csarron/bert-base-uncased-squad-v1

28. https://huggingface.co/ozcangundes/T5-base-for-BioQA

29. Velickovic, P., Cucurull, G., Casanova, A., Romero, A., Lio, P., Bengio, Y., et al.: Graph attention networks. Stat **1050**(20), 10–48550 (2017)

30. Mrini, K., Farcas, E., Nakashole, N.: Recursive tree-structured self-attention for answer sentence selection. In: Proceedings of the 59th Annual Meeting of the Association for Computational Linguistics and the 11th International Joint Conference on Natural Language Processing (Volume 1: Long Papers), pp. 4651–4661 (2021)

Tropical Cyclone Analysis and Accumulated Precipitation Predictive Model Using Regression Machine Learning Algorithm

Maribel S. Abalos[1]([⊠]) and Arnel C. Fajardo[2]

[1] College of Engineering Architecture and Technology, Isabela State University, Ilagan, Isabela, Philippines
maribel.s.abalos@isu.edu.ph
[2] College of Computing Studies, Information and Communication Technology, Isabela State University, Cauyan City, Isabela, Philippines

Abstract. The Philippines is susceptible to tropical cyclones due to its geographical location which generally produces heavy rains and flooding of a large area that result in heavy casualties to human life and destruction of crops and properties. In the past 4 years, the Cagayan Valley region is hit by almost 20 Tropical Cyclones with different strengths and attributes. It is of utmost importance to have sufficient knowledge of such maritime phenomena for beneficial purposes. This study analyzed the Tropical Cyclones that hit Cagayan Valley using Philippine Atmospheric Geophysical and Astronomical Services Administration (PAGASA) dataset from 2017–2021. Different attributes were considered to create a framework, statistical analysis, and comparison matrix. Also, the dataset is analyzed using four Regression Machine Learning Algorithms namely MultilayerPerceptron, Linear Regression, SMOReg, and Gaussian Processes, and compared their performances where the best-performed algorithm is used to develop a predictive model for rainfall accumulation. Results of the study found that Cagayan Valley accumulated 219.857 mm of rainfall and an average Tropical Cyclone Maximum Sustained winds of 123.438 kph. Tropical Cyclone Ulysses reached almost the highest value in all the attributes as evidence of its devastating nature last November 2020. Also, the Linear Regression algorithm revealed a better performance which displayed a higher correlation coefficient and lower error estimates and took less time in developing the model. Lastly, the results can be considered to be a basis of the impact of a Tropical Cyclone based on their attributes and can be implemented into technological platforms for predictive application which helps foresee the prediction rate of rain accumulation to develop techniques and preventive measures to target reduction of damage on properties and the number of casualties.

Keywords: Tropical Cyclone Analysis · Predictive Model · Regression Machine Learning Algorithm

© The Author(s), under exclusive license to Springer Nature Switzerland AG 2023
F. Neri et al. (Eds.): CCCE 2023, CCIS 1823, pp. 203–219, 2023.
https://doi.org/10.1007/978-3-031-35299-7_17

1 Introduction

One of the world's most susceptible countries to natural hazards from geographical to meteorological aspects is the Philippines which suffers more natural hazards like earthquakes, volcanic eruptions, typhoons, floods, droughts, and landslides than any other country with an average of eight disasters per year. About 20 typhoons each year, an equivalent of 25% of the global occurrences of typhoons in the Philippine Area of Responsibility [1] – third in disaster risk among all countries [2]. From 2000 to 2019, 304 disaster events happened in the Philippines and an average of 7796 people per 100000 population were affected [3]. These number of disasters implied a frequent natural hazard that affect the country. Philippine Atmospheric Geophysical and Astronomical Services Administration PAGASA, describes the climate of the Philippines as tropical and maritime – characterized by relatively high temperature, high humidity, and abundant rainfall [4].

The administration emphasized that rainfall is the most climatic element in the Philippines. PAGASA pointed out that rainfall associated with tropical cyclones is both beneficial and harmful. Although the rains contribute to the water needs of the areas traversed by the cyclones, the rains are harmful when the amount is so large as to cause flooding. Rainfall distribution throughout the country varies from one region to another, depending upon the direction of the moisture-bearing winds and the location of the mountain systems. The country's annual rainfall based on available data of PAGASA, varies from 965 to 4,4064 mm annually. One of the natural disasters that produce an abundant amount of rainfall is Tropical Cyclones or typhoons which is due to their heavy precipitation. Warm, humid air associated with tropical cyclones produces a very large amount of rainfall, often in excess of 200 mm in just a few hours. This can cause short-term flash flooding, as well as slower river flooding as the cyclone moves inland [4]. Of all the meteorological disasters, tropical cyclones impose significant threats to human lives and urban development because of their high frequency of occurrence and massive destructive power that leads to inundation caused by storm surges and heavy precipitation [5].

Region 02 or Cagayan Valley in the Philippines is considered a catch basin of rainwater from surrounding mountain ranges that cause the rivers to overflow, prompting massive flooding - in the wake of Typhoon Ulysses. Previous tropical cyclones had also contributed to the destructive power of the typhoon. Based on the report by Climate Change Commission, typhoon Ulysses carries 356 mm of rain for a day during its path in the region. Affected 5 million persons from different regions – almost 12 billion in the total cost of infrastructure damage and 7 billion in agricultural damage. The typhoon caused incidents like 213 flooded areas. This catastrophic incident led to the loss of lives and properties. Different typhoons had passed the region for several years leading to the same calamity and situation for the people living in the area. Damages and losses increase dramatically. However, research studies highlight that there still remains a limited understanding of the factors contributing to damage and loss. There is also a paucity of studies mapping disaster risk reduction at regional levels, with a need for research conducted to better understand the impact of disasters in an area [6].

The country's National Disaster Risk Reduction and Management Plan 2011–2018, pushed a framework of safer, adaptive, and disaster-resilient Filipino communities toward sustainable development. One of its thematic areas is Disaster Preparedness which establishes and strengthens the capacities of communities to anticipate the negative impacts of emergency occurrences and disasters [7]. With the help of innovative technology nowadays in high-powered specification of computers, early warning signs in the anticipation of disasters is possible through the development of a predictive model using historical data of tropical cyclones.

Many researchers have used artificial neural networks (ANNs), model classifiers and vector machines to analyze traditional classification and prediction problems among large amounts of datasets [8–10]. ANNs provide a compact method of considering large amounts of data and a simple means of assessing the likely outcome of a complex problem with a specified set of conditions. ANN is used for several applications including the prediction of historical data. PAGASA has been tracking and recording tropical cyclones for several years. A typhoon or tropical cyclone is characterized based on the PAGASA report, maximum sustained winds (msw), gustiness, mean surface level pressure (mslp), and rain accumulation. These recorded data can be used for the implementation of Artificial Neural Network and Machine Learning Algorithms for analysis and prediction.

Calamity or disaster records are kept in the database of a particular system. Throughout the year they are accumulated and produced a large amount of data with similar entities. Using machine learning or artificial neural network, these datasets can be analyzed to provide important insight or patterns or a way to discover a possible solution for a problem. Prediction or forecasting the possible danger of a natural phenomenon is important for the information of the people that it provides the early warning signs of what might happen that will lead them to prepare for safety. PAGASA recorded many tropical cyclones in Cagayan Valley that led to not a few but so many casualties due to the unprepared or late analysis of the incoming disaster. Since Cagayan Valley is considered a catch basin due to mountain ranges, the rainfall amount of typhoons contributes to the rise of The Cagayan River causing floods in nearby areas or low-lying places.

This study focuses on the analysis of accumulated precipitation of different typhoons that landed in the Cagayan Valley over the past 5 years. Also, to develop a model for predicting the likelihood of the amount of accumulated precipitation for immediate action before the landfall of the typhoon.

2 Related Literature

2.1 Philippine as Disaster Prone Country

The Philippines is one of the most vulnerable countries in the world based on several global disaster risk reports - natural extreme weather events. It lies in the middle of the Pacific Typhoon Belt, which includes some of the countries most exposed to tropical storms and other natural disasters, such as floods, droughts, landslides, earthquakes, tsunamis, and volcanic eruptions. On average, 20 tropical cyclones every year enter or develop within the Philippine Area of Responsibility, the highest in the world, with 8–9 cyclones making landfall.2 Several climate vulnerability rankings where the country is consistently at the top affirm the vulnerability of the Philippines. The 2022 World Risk Report [11] describes the Philippines as especially vulnerable to climate insecurity with extreme weather events [12].

During the past decade, extreme weather events, particularly super typhoons and intense monsoon rains, hit the Philippines with regularity. Typhoon Yolanda (Haiyan), which caused an estimated P571.1 billion in damages and losses (World Bank, 2017) along with thousands of casualties, easily comes to mind. The unprecedented scale and magnitude of the impact of Typhoon Yolanda made the Philippine government take a second look at the appropriateness of its disaster-resiliency and rehabilitation programs that can mitigate the impact of such extreme weather events on the economy and on the welfare of the people. Typhoon Ulysses has sustained winds of 118e220 km/hand 64e119 knots. Based on the report of the Climate Change Commission, Ulysses carries 356 mm of rain for the whole day of November 12. About 5,184,824 persons from Region 1e3, CALABARZON, MIMAROPA, Region V, CAR and NCR were affected. In terms of casualties, there were 101 dead, 85 injured, and 10 missings [5]. Rainfall remains one of the most influential meteorological parameters in many aspects of our daily lives. With effects ranging from damage to infrastructure in the event of a flood to disruptions in the transport network, the socio-economic impacts of rainfall are noteworthy [13, 14]. Floods and similar extreme events are consequences of climate change that are expected to occur more frequently and have catastrophic effects in years to come [15]. In addition to the cost in terms of forgone output, productivity losses, and fiscal and external sustainability, extreme weather events pose risks to the soundness of financial institutions and the stability of the overall financial system [16].

2.2 Early Warning Sign and Predictive Model

Rainfall forecasting has gained utmost research relevance in recent times due to its complexities and persistent applications such as flood forecasting and monitoring of pollutant concentration levels, among others. Existing models use complex statistical models that are often too costly, both computationally and budgetary, or are not applied to downstream applications. Therefore, approaches that use Machine Learning algorithms in conjunction with time-series data are being explored as an alternative to overcome these drawbacks [15].

The use of Artificial Neural Networks (ANNs) as rainfall forecasting models has widely captured the attention of researchers [16–18]. One factor that played an important role in the widespread use of ANNs in this field of application was the emergence of wireless technologies, such as the Internet of Things, which accelerated the development of inexpensive and effective solutions for capturing satellite imagery and historical radar data. Another relevant factor that contributed to the growth of the use of ANNs as an approach to forecast rainfall is its capability to address non-linearity in rainfall data and that they require little or no knowledge of the relationships between the variables that are being considered. Caruana et al. have shown that a data-driven approach using a machine learning method is capable of establishing ahead-of-time flood predictions and is a viable approach when there is a lack of a geological and hydrological model of a specific site [19]. This study [20, 21] set out to compare the prediction performance of rainfall forecasting models based on LSTM-Networks architectures with modern Machine Learning algorithms.

3   Methodology

3.1   Experimental Methodology

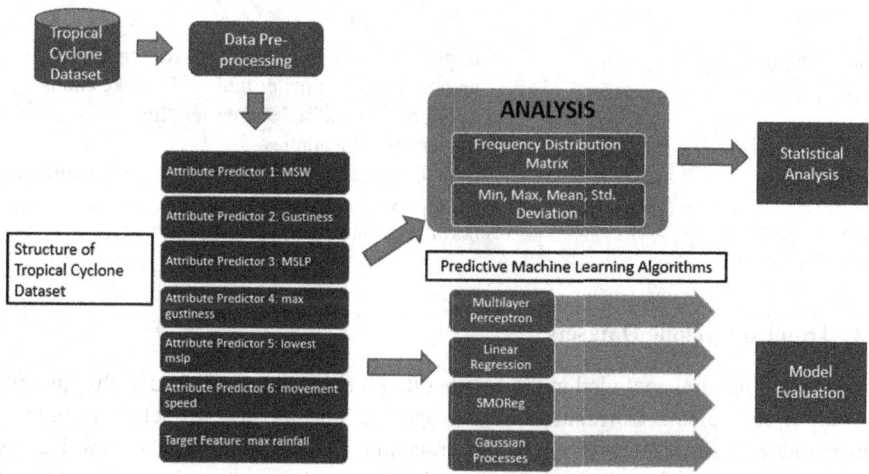

Fig. 1. Experimental methodology.

This study followed the Experimental Methodology as depicted in Fig. 1. The Dataset retrieved from PAGASA underwent several stages – data preprocessing takes the first step such as data cleaning and data replenishment. Selected data is reflected after data preprocessing which are used for analysis. Different predictive machine learning algorithms are implemented to the processed dataset which then evaluated as predictive model.

Shown in Table 1 the description of different regression machine learning algorithms. There were 4 algorithms implemented – Multilayer Perceptron, Linear Regression, SMOReg and Gaussian Processes. These algorithms were used and evaluated their performance on the dataset.

Table 1. Description of regression machine learning algorithms.

Regression Machine Learning Algorithm	Description
Multilayer Perceptron	A classifier that uses backpropagation to learn a multi-layer perceptron to classify instances
Linear Regression	Uses the Akaike criterion for model selection, and is able to deal with weighted instances

(continued)

Table 1. (continued)

Regression Machine Learning Algorithm	Description
SMOReg	Implements the support vector machine for regression. The parameters can be learned using various algorithms
Gaussian Processes	Implements Gaussian processes for regression without hyperparameter-tuning. To make choosing an appropriate noise level easier, this implementation applies normalization/standardization to the target attribute as well as the other attributes (if normalization/standardizaton is turned on)

3.2 Tropical Cyclone Datasets

The climatic features included in the weather datasets are first described. Subsequently, the description of the Correlation Matrix analysis and the feature selection process carried out as part of the pre-processing procedure to prepare the time-series data for use in the training of the rainfall forecast models is given. The rationale for developing a prediction model lies in the variables of the given dataset. The dataset is retrieved from PAGASA recorded Tropical Cyclone that landed in the Cagayan Valley region from the year 2017–2021. The original variables in the dataset are shown in Fig. 2.

Fig. 2. Tropical cyclone dataset attributes.

The dataset contains variables viable for testing and analysis. The given variables point from Tropical Cyclone Information to Information Track. Many of the select attributes are merged in one of the entities and extracted them for consolidation.

3.3 Data Pre-processing

Historical data is used for analysis in determining the predictive accuracy of the dataset. Attributes were selected using a two-way step process, subset generation, and ranking [12] in order to address the objectives of this study.

As with any Machine Learning approach in Artificial Neural Networks, a processing procedure is required to prepare raw data for use in model training and testing processes. The pre-processing procedure carried out had as its objective the determination of significant attributes by eliminating unnecessary attributes, the deletion or replenishment of incomplete data, and averaging multiple possible values that suffice a single value.

Specifically, the pre-processing procedure performed is described in Fig. 3:

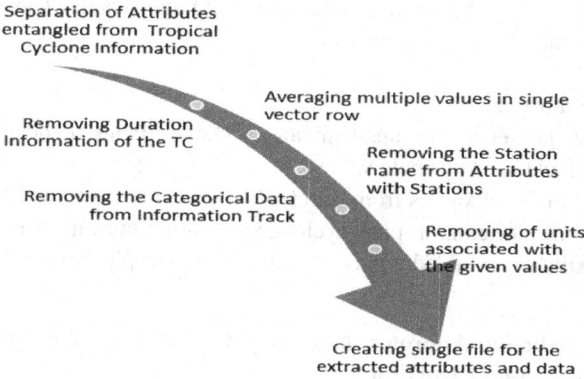

Fig. 3. Data-preprocessing procedures.

- Separation of Attributes entangled from Tropical Cyclone Information. This step aims to detached details from Tropical Cyclone Information to create different columns for such details. Such as Name and Gustiness
- Removing Duration Information of the Tropical Cyclone. The duration which contains Date and Time of the Tropical Cyclone which is apparently not significant for creating the model.
- Averaging multiple values in single vector row. This part requires averaging calculations as some attributes reflect multiple values for different stations.
- Removing the Station name from Attributes with Stations. This is to delete station names associated with the attributes that show stations and a value.
- Removing of Units associated with the given values. Values are in numerical form with specified units which do not give so much significance prior to testing and evaluation.
- Removing the Categorical Data form Information track. This is to remove the entire Information Track Attribute for it contains categorical data which are not needed in the study.
- Creating Single File for the Extracted Attributes and Data. The remaining attributes is extracted together with its values and save as separate file for testing, evaluation, training and development of a predictive model.

3.4   Data Mining Tool

This study implemented the use of Excel for Data Preprocessing and Weka for dataset testing and evaluation.

Statistical Analysis

The dataset is subjected to statistical analysis in determining the minimum, maximum, mean, standard deviation and range. Prior analytical description is written based on the value reflected on each of the criterion and how it is associated with the variability and dispersion of the dataset such as standard deviation. The minimum value is taken from the lowest value of a certain attribute while the maximum value refers to the highest value. The difference between the maximum and minimum value is the range and is calculated by subtracting the smallest value from the largest value. This is to determine the variability of the values of the dataset.

Statistical Distribution

Values from the dataset are grouped in ranges per attributes to determine Tropical Cyclones' classified based attribute values.

The grouped range of values from different attributes are analyzed and interpreted. This is to determine particular tropical cyclone has dominating values among the others. Such as the maximum sustained winds or gustiness of a tropical cyclone.

3.5   Rainfall Prediction Approach and Prediction Model Using Regression Machine Learning Algorithm

Using Weka as a tool for data analysis, different machine learning algorithms were used in the evaluation of the dataset. The dataset is evaluated based on Split Percentage and 10-Cross Fold Validation.

Multi-layer perceptron (MLP) is a supplement of a feed-forward neural network. It consists of three types of layers—the input layer, the output layer as shown in Fig. 4 and the hidden layer, as shown in Fig. 3 Multilayer Perceptron.

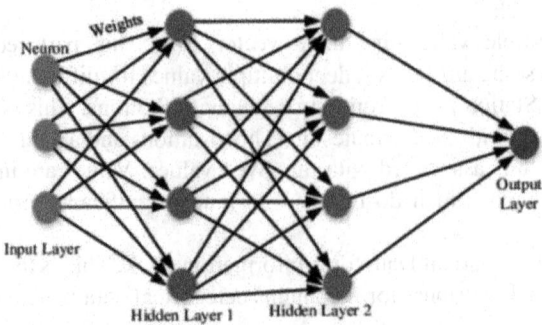

Fig. 4.   Schematic representation of MLP.

SMOReg

The SMOreg algorithm is a useful machine learning (ML) algorithm for analyzing both categorical and numerical data sets by allowing the acquired model to be analyzed in terms of the weight of each attribute and, therefore, the significance of each attribute can be understood;

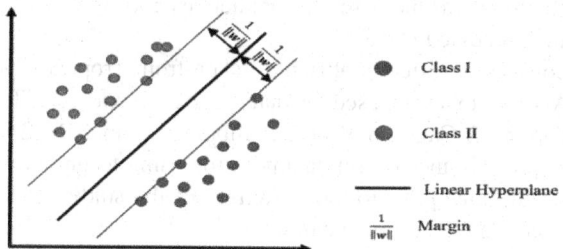

Fig. 5. Schematic of classification of SMOReg.

Figure 5 illustrates the classified data with a linear hyperplane separating the two classes, shown in blue and red. Maximizing the margin is equivalent to maximizing the term $1/\|w\|$ which is equivalent to minimizing $\|w\|$. The optimization of the hyperplane fitting is achieved through variational calculus and Lagrangian methods.

Gaussian Processes

A Gaussian process is a <u>random process</u> where any point $\mathbf{x} \in \mathbb{R}^d$ is assigned a random variable $f(x)f(x)$ and where the joint distribution of a finite number of these variables $p(f(x1),...,f(xN))p(f(x1),...,f(xN))$ is itself Gaussian:

For this study, the prior GP needs to be specified. The prior mean is assumed to be constant and zero or the training data's mean. The prior's covariance is specified by passing a kernel object. The hyperparameters of the kernel are optimized during the fitting of GaussianProcessRegressor by maximizing the log-marginal-likelihood (LML) based on the passed optimizer. The first run is always conducted starting from the initial hyperparameter values of the kernel; subsequent runs are conducted from hyperparameter values that have been chosen randomly from the range of allowed values. If the initial hyperparameters should be kept fixed, none can be passed as an optimizer.

Predictive Model

Based on the evaluation of the algorithms on the particular criteria, model evaluation was created as a matrix of comparison. The dataset was subjected to implement different machine-learning algorithms as explained and mentioned earlier in this paper. Using Weka, the processed dataset is loaded in the Weka Explorer tool then chose the class for prediction, and saves the developed model for future use for forecasting. The rain accumulation attribute of the dataset is the class to be predicted.

4   Results and Discussion

With the steps involved in processing the data for analysis, results were obtained and tabulated which yielded different frequency distribution tables and presented using a graphical presentation. The dataset was used to compare and evaluate the performance of regression machine learning algorithms in developing a model for predictability. The best-performed classifier was then used for predictive analysis.

Structure of the processed Dataset.

The dataset consists of different attributes taken from Tropical Cyclones recorded information of PAGASA that were used for analysis. It is composed of Tropical Cyclones that landed in Region 02 Cagayan Valley, Philippines from 2016-2A two-wayo-way attribute selection process and several data pre-processing techniques were utilized to obtain significant attributes prior to the objectives of the study. Table 2 is the list of attributes after processing the original dataset.

Table 2. Determined attributes of the processed dataset.

Attribute	Description
tc (Ompong, Egay, Falcon, Ineng, Jenny, Hanna, Marilyn, Onyok, Quiel, Ramon, Tisoy, Pepito, Rolly, Siony, Tonyo, Ulysses)	Tropical Cyclone: Pertains to the Name of the tropical cyclones
msw_kph	Maximum Sustained Winds in kilometer per hour
gust_hpa	Gustiness in Hectopascal
mslp_hpa	Minimum Sea level pressure in Hectopascal
max_gust_kph	Maximum Gustiness in kilometer per hour
lowest_mslp_hpa	Lowest mean Surface Level Pressure in Hectopascal
mov_speed_kph	Movement Speed in Kilometer per hour
max_rainfall_accu	Maximum rainfall accumulation in Millimeters

Figure 6 shows Tropical Cyclones' maximum sustained winds. TC Rolly had the highest maximum sustained with 225 km per hour which implied strong winds and classified as Super Typhoon that caused huge damage to properties. Also, TC Ompong, Hanna, Tisoy, and Ulysses carried the top for having the highest msw. During the duration of these Tropical Cyclones, many had recorded high-cost damage to properties and loss of lives.

TC Ompong and Rolly had the highest gustiness which ranges from 250–350 km per hour as peak wind speed as shown in Fig. 7 but TC Ompong had the highest maximum gustiness depicted in Fig. 8 which implies that these TCs have the highest intensity in terms of wind gustiness.

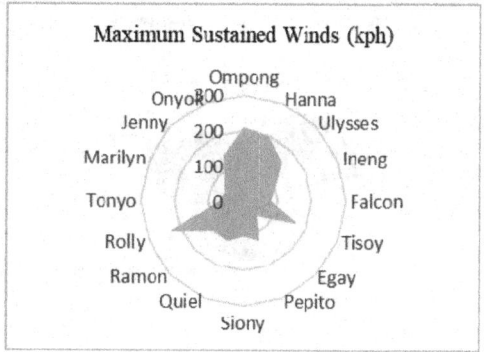

Fig. 6. Maximum sustained winds of the tropical cyclones in kilometer per hour.

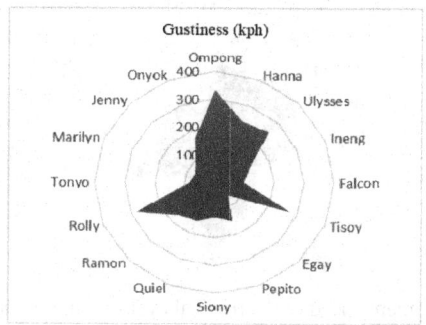

Fig. 7. Gustiness of the tropical cyclones in kilometer per hour.

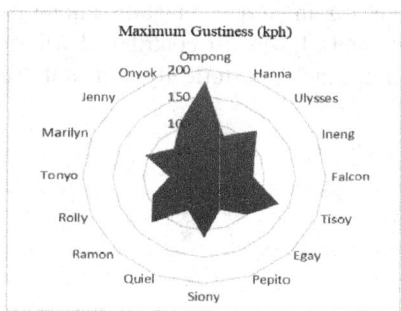

Fig. 8. Maximum gustiness of the tropical cyclones in kilometer per hour.

Among the Tropical Cyclones as shown in Fig. 9, TC Egay, Tonyo and Jenny have the highest Minimum Sea Level Pressure i.e. ranging from 995 to 1004 Hectopascal.

From Fig. 10, the fastest Tropical Cyclone based on the available data gathered were TC Siony and Falcon with 30 kph. While TC Hanna and Ramon were the slowest. This means that these TCs had a longer duration (movement speed) in the region.

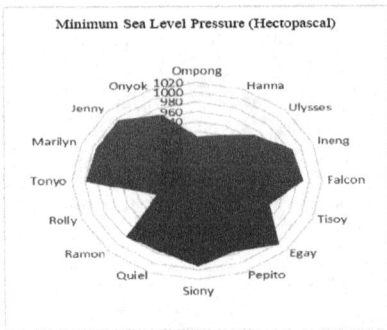

Fig. 9. Minimum sea level pressure of the tropical cyclones in hectopascal.

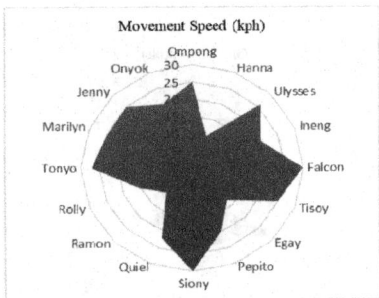

Fig. 10. Movement speed of the tropical cyclones in kilometer per hour.

With 427.5 mm of rain accumulation, TC Ompong caused flash floods in Cagayan Valley which was the highest rainfall accumulation among the Tropical Cyclones as shown in Fig. 11. TC Hanna and Ulysses also contributed to the top rainfall accumulation ranging from 330 to 370 mm which is historically evident that these TCs devastated the region for flash floods.

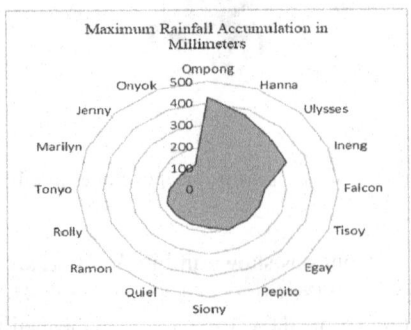

Fig. 11. Rain accumulation of the tropical cyclones in millimeters.

Among all the attributes for Tropical Cyclones, TC Ulysses reached almost the highest value in all the categories. This is evident, in November 2020 when TC Ulysses hit the region which left high damage to properties and many casualties.

Minimum, Maximum, Mean, Standard Deviation and Range

The processed dataset is subjected to statistical analysis, determining its minimum, maximum, mean, standard deviation, and range of each of the selected attributes. Among all the tropical cyclones that landed, it is shown in Table 3, that Cayagan Valley received 427.5 mm of rain accumulation which 30 kph highest movement speed of tropical cyclones landed specifically in the duration of TC Ompong.

Table 3. Results of the analysis on minimum, maximum, mean, standard deviation and range for the processed dataset.

	msw_kph	gust_kph	mslp_hpa	max_gust_kph	lowest_mslp_hpa	mov_speed_kph	max_rainfall_accu
Min	55	65	890	56	916.1	10	123
Max	225	330	1004	178.4	1003.65	30	427.5
Mean	123.438	167.188	966.875	99.471	979.95	21.438	219.857
Std. Deviation	53.688	83.725	34.531	30.792	23.441	6.48	93.197
Range	175	265	114	122.4	87.55	20	304

In general Cagayan Valley accumulated 219.857 mm of rainfall and an average TC Maximum Sustained winds of 123.438 kph. The mean of the Tropical Cyclones for movement speed in the region was 21.438 kph with 99.471 average maximum gustiness.

Implementation and Comparison of Regression Machine Learning Algorithms in the Dataset

Further analysis is performed on the dataset based on the time the model was built. In Table 4, regression algorithms are compared on how fast the model was built. By building with 5 times execution rate in the same classifier then getting the average time taken. The Model Build Time parameter for comparison was executed in the same software (Weka) and computer specifications. It is then presented that MultilayerPerceptron took less time in developing the model followed by the Gaussian Processes algorithm.

Table 4. Use training set: Classifier execution time.

ML Algorithm	Model Build Time (seconds)
MultilayerPerceptron	0.01
Linear Regression	0.001
SMOReg	0.03
Gaussian Processes	0.01

10-Fold Cross Validation Test

To summarize the overall result of the performance of each of the Machine Learning algorithms, Fig. 12 shows the evaluation of classifiers taken from the output result Weka. It tested with Paired T-Tester (corrected) with significance of 95%. The regressions and the true classes in a chance corrected measure shows that Linear Regression classifier performed better than other classifiers with 0.02 correlation coefficient which implies a high positive correlation. It is measured using 10-fold cross validation. Furthermore, mean absolute error measures the accuracy of the variable's continuity which in line to Root Mean Squared Error that both used for the diagnosis of variation errors in a set of forecasts. In Fig. 12, Multilayer Mean Absolute error and Root Mean Squared Error difference is 63.04, therefore, JRip has the greater variance in the individual errors in the sample followed by Linear Gaussian Processes and Linear Regression algorithms. By means of relative absolute error and relative root squared error, Linear Regression showed lowest values less than 100.

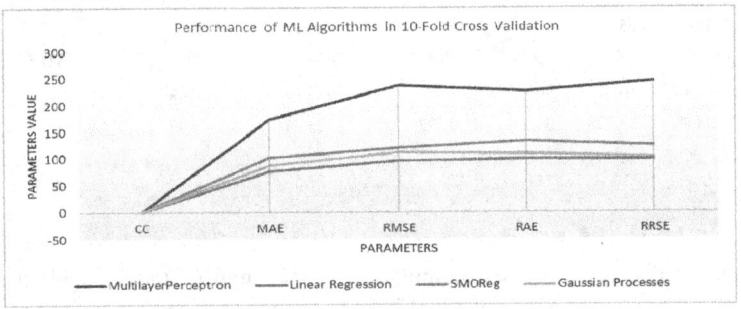

Fig. 12. Summary evaluation of classifiers' performances using 10-fold cross validation.

Split Percentage: 66/34

The model is produced using training data. And the data set with known output values are used in developing the model. However, the model type has an entire training set and is divided by 34/66 ratio. 34% for a training data set which we used for developing the model and 66% for the test data which was used for testing the accuracy of the model in comparison to different ML algorithms.

The graphical representation of the equivalent classifier output shown in Fig. 13 demonstrates the comparison of their keen differences. SMOReg algorithm has a correlation coefficient of 0.15 followed by Linear Regression of 0 other algorithms fall at negative correlation values. Just a difference of less than 1 point Linear Regression has the smallest Mean Absolute value compared to SMOreg. Also, Linear Regression's Root Mean Squared Error is 6 less than SMOReg and made to the smallest value of RMSE. Linear Regression also has the second smallest Relative Absolute error Value i.e. 100 which is the same to it RRSE.

Predictive Model using Linear Regression Algorithm

By performance, the Linear Regression ML algorithm showed exceptional work with a

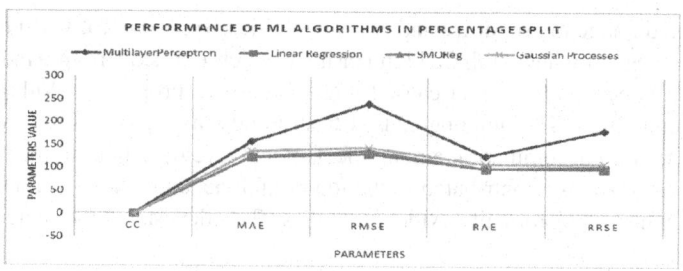

Fig. 13. Classifier algorithms: Split percentage test result.

higher positive correlation coefficient and lower error estimates which were then used as a predictive algorithm for rain precipitation.

Table 5 assumed numerical value for a certain Tropical Cyclone attribute that may pass in the Region. Using the developed model, it is predicted in the first row of 335.045 mm of rainfall precipitation given the data from msw_kph to lowest_mslp_hpa. Also, 160.409 mm of rainfall accumulation is predicted from the second row of data. The error estimates of the algorithm are relative to the dataset which is good enough for the model to be dependent on for predicting the Accumulated Precipitation of a Tropical Cyclone.

Table 5. Test data for prediction.

msw_kph	gust_kph	mslp_hpa	max_gust_kph	lowest_mslp_hpa	max_rainfall_accu	Predicted Value
124	340	998	165	879	?	335.045
113	78	879	234	986	?	160.409

5 Conclusion

Tropical Cyclones are natural phenomena that may bring disaster and leave quite a damage and loss of lives. It contributes to major effects on the life of individuals and the economic growth of a country. Analysis of Tropical Cyclones with predictive model gained more attention nowadays for climate change and the occurrence of Tropical Cyclones is unexpected.

This study provided a systematic framework under quantitative techniques to analyze and graphically interpret the Tropical Cyclones that hit Region 2 or Cagayan Valley in the Philippines. The attributes/characteristics of the Tropical Cyclones were used trimmed down into significant attributes for use as input tools and subjected to analysis. It is found that Cagayan Valley accumulated 219.857 mm of rainfall and an average TC Maximum Sustained winds of 123.438 kph. Tropical Cyclone Ulysses reached almost the highest value in all the attributes. As evident, in November 2020 when TC Ulysses hit the region that left high damage to properties and many casualties. Typhoon Rolly had the highest maximum sustained winds with 225 km per hour which implied strong winds and was classified as Super Typhoon that caused huge damage to properties. With 427.5 mm of rain accumulation, TC Ompong caused flash floods in Cagayan Valley which was the highest rainfall accumulation among the Tropical Cyclones that hit the region. TC Hanna and Ulysses also contributed to the top rainfall accumulation ranging from 330 to 370 mm which is historically evident that these TCs devastated the region for flash floods.

The results of this study are considered to be a basis of how Tropical Cyclones impact a certain area or region based on their characteristics/attributes. Furthermore, the findings of the study showed that Cagayan Valley as a catch basin in the North of the Philippines requires attention from the government for rehabilitation programs of the settlers in low-lying areas. In every Tropical Cyclones that passed the region, there is still high rain accumulation which is one of the factors of the suffering of the people for damage to properties and a high count of casualties is always in the depth of expectation.

Lastly, the machine learning algorithms used in developing a model for predictive analysis on the supplied dataset delivered a comparative study in which Linear Regression revealed a better performance which displayed a higher correlation coefficient and lower error, and took less time in developing the model. This result can be accommodated and implemented into mobile or web-based platforms for predictive application, various agencies and other organizations which helps them foresee the prediction rate of rain accumulation and develop techniques, preventive measures, and plans to target minimization of damage to properties and number of casualties from a Tropical Cyclone.

References

1. Lloyd, S., Gray, J., Healey, S., Opdyke, A.: Social vulnerability to natural hazards in the Philippines. Int. J. Disaster Risk Reduct. (2022). https://doi.org/10.1016/j.ijdrr.2022.103103
2. World Economic Forum The Global Risks Report 2018 (13th edn. en) (2019) Technical report OCLC: 1099890423. http://www3.weforum.org/docs/WEF_Global_Risks_Report_2019.pdf (visited on 10/26/2021) Google Scholar
3. UNDRR The Human Cost of Disasters - An overview of the Last 20 Years 2000–2019 (2019) en. Technical report. https://reliefweb.int/report/world/human-cost-disasters-overview-last-20-years-2000-2019 (visited on 03/31/2021) Google Scholar
4. PAGASA Website. https://www.bbc.co.uk/bitesize/guides/z9whg82/revision/2
5. Santos, G.D.C.: 2020 tropical cyclones in the Philippines: a review. Tropical Cyclone Res. Rev. 10(3), 191–199 (2021). https://doi.org/10.1016/j.tcrr.2021.09.003
6. Yonson, R., Noy, I., Gaillard, J.C.: The measurement of disaster risk: an example from tropical cyclones in the Philippines. Rev. Dev. Econ. 22(2), 1467–9361 (2018). pp. 736–765, https://doi.org/10.1111/rode.12365. eprint: https://onlinelibrary.wiley.com/doi/abs/10.1111/rode.12365 (visited on 03/12/2021)
7. National Disaster Risk Reduction and Management Plan (NDRRMP) (2011)
8. Hong, J.: An improved prediction model based on fuzzy-rough set neural network. Int. J. Comput. Theory Eng. 3(1), 158–162 (2011)
9. Owramipur, F., Eskandarian, P., Mozneb, F.S.: Football result prediction with Bayesian network in Spanish league-Barcelona team. Int. J. Comput. Theory Eng. 5(5), 812–815 (2013)

10. Yusof, Y., Mustaffa, Z.: Dengue outbreak prediction: a least squares support vector machines approach. Int. J. Comput. Theory Eng. **3**(4), 489–493 (2011)
11. Heintze, H.-J., et al.: World Risk Report 2018 Bündnis Entwicklung Hilft and Bochum: Institute for International Law of Peace and Armed Conflict. Ruhr University Bochum, Aachen (2018)
12. Duncan, A., Keedwell, E., Djordjevic, S., Savic, D.: Machine learning based early warning system for urban flood management. In: ICFR 2013: International Conference on Flood Resilience: Experiences in Asia and Europe, University of Exeter, UK, 5–7 September 2013
13. Singh, P., Borah, B.: Indian summer monsoon rainfall prediction using artificial neural network. Stoch. Env. Res. Risk Assess. **27**(7), 1585–1599 (2013)
14. Chakrabarty, H., Murthy, C.A., Gupta, A.D.: Application of pattern recognition techniques to predict severe thunderstorms. Int. J. Comput. Theory Eng. **5**(6), 850–855 (2013)
15. Barrera-animas, A.Y., et al.: Machine learning with applications rainfall prediction: a comparative analysis of modern machine learning algorithms for time-series forecasting. Mach. Learn. Appl. **7**(August 2021), 100204 (2022). https://doi.org/10.1016/j.mlwa.2021.100204
16. Gnanambal, D., Thangaraj, D., Meenatchi, V.T., Gayathri, D.: Classification algorithms with attribute selection: an evaluation study using WEKA. Int. J. Adv. Network. Appl. **9**(6), 3640–3644 (2018). http://oaji.net/pdf.html?n=2017/2698-1528114152.pdf
17. Singh, P., Pal, G.K., Gangvar, S.: Prediction of cardiovascular disease using feature selection techniques. Int. J. Comput. Theory Eng. **3**, 97–103
18. Thamilselvan, P.: Lung cancer prediction and classification using AdaBoost data mining algorithm. Int. J. Comput. Theory Eng. **14**(4), 149–154 (2022)
19. Liu, Q., Zou, Y., Liu, X., Linge, N.: A survey on rainfall forecasting using artificial neural network. Int. J. Embedded Syst. **11**(2), 240–249 (2019)
20. Radhika, Y., Shashi, M.: Atmospheric temperature prediction using support vector machines. Int. J. Comput. Theory Eng. **1**(1), 55–58 (2009)
21. Chitra, A., Uma, S.: An ensemble model of multiple classifiers for time series prediction. Int. J. Comput. Theory Eng. **2**(3), 454–458 (2010)

Mapping Learning Algorithms on Data, a Promising Novel Methodology to Compare Learning Algorithms

Filippo Neri[✉][iD]

DIETI, University of Naples, via Claudio 21, 80100 Naples, Italy
filippo.neri.email@gmail.com

Abstract. The paper describes a novel methodology to compare learning algorithms by exploiting their performance maps. A performance map enhances the comparison of a learner across learning contexts and it also provides insights in the distribution of a learners' performances across its parameter space. Also some initial empirical findings are commented.

In order to explain the novel comparison methodology, this study introduces the notions of learning context, performance map, and high performance function. These concepts are then applied to a variety of learning contexts to show how the methodology can be applied.

Finally, we will use meta-optimization as an instrument to improve the efficiency of the parameter space search with respect to its complete enumeration. But, note that meta-optimization is neither an essential part of our methodology nor the focus of our study.

Keywords: Learning algorithms · Comparing Learners · Performance maps

1 Introduction

The standard approach used in machine learning to compare learning algorithms consists in contrasting their performances on a data set unseen during the learning phase. A learner's performance is expressed in the form of a single numeric value representing, for instance, its accuracy, the error rate, etc. Usually a confidence interval around the mean performance value is also provided. However, in the end, a whole learner behavior is condensed into just one single number (i.e. the mean accuracy). All other information about the learning process (i.e. how the search in the hypothesis space was conducted, what effect changing learning parameters produces, how human readable is the found concept, etc.) is simply discarded. From the theoretical point of view, the user is then supposed to select a learner over the other just by considering a single number.

On the opposite, from the practical point of view, the literature papers may only partially helpful as they usually hide away the important step of *parameter selection* that is, however, performed by the authors but generally not discussed in the paper.

F. Neri et al. (Eds.): CCCE 2023, CCIS 1823, pp. 220–231, 2023.
https://doi.org/10.1007/978-3-031-35299-7_18

When considering real data, we believe, instead, that a) the step of parameter selection should be considered a full part of the learning process, and that b) a learner's parameter sensitivity should play a role in comparing learners across different learning contexts. In fact, if a learner's result is very sensitive to its settings, the user may want to consider selecting a lower performing learner with stabler results to ensure a more robust behavior on future data.

Following the above considerations, this study describes a new methodology to compare learning systems by using *performance maps* that makes explicit a learner's sensitivity to its parameter settings.

We define a *performance map* as the set of performance values, associated to the parameter settings that produced them, when a leaner is applied to some data. *Performance maps* are functions of *learning contexts*. In order to understand how to build a performance map, let us then define what a *learning context* is for the extent of this study.

A *learning context LC* is a quadruple made of:

1. a learning algorithm L,
2. a parameter space search method M (grid search or a meta-optimizer to reduce the run time),
3. the subset of the parameter space explored by the search method $MOPS$ (the set of parameter settings for L considered during the search) and
4. a data set D.

In the following, we will use the term meta-optimizer to identify the parameter space search method in a learning context. The meta-optimization method M can take the form of a simple grid search (which usually takes a long time to run) or a faster search algorithm. We recall that meta-optimizing a learner consists in finding the best performing parameter settings for the learner by partially searching the space of all its possible parameter settings [2,7,8,11,12,33]. Meta-optimization can significantly improve the performance of a learning algorithm [4,5].

Then meta-optimization of a learner L is accomplished by performing multiple runs of L on D, using several parameter settings, in order to evaluate L's performance for each considered parameter settings. Either exhaustive search or a specific meta-optimization algorithm M can be used. And, the set of L's parameter settings evaluated during the meta-optimization process is the *meta-optimized parameter space* $(MOPS)$. The collection of pairs $< s, L$'s performance $>$, with s in $MOPS$, allows to create the *performance map(LC)* that we are interested in.

The selected meta-optimization method M determines the composition of $MOPS$ and, in turns, of the *performance map(LC)*. Performance maps can be either complete, if $MOPS$ is equal to the set of all parameter settings for L, or partial/approximated, if $MOPS$ is a proper subset of it.

Moreover, we stress the point that meta-optimization is not the focus of this research, we will simply use it as a tool to build performance maps in a more efficient way with respect to the complete enumeration of the parameter space.

Finally, in the description of how performance maps are created, the machine learner expert can easily recognize a formalized version of the manual parameter tuning process accomplished by all authors in order to select the 'most suitable' configuration for running the learners discussed in their papers.

Novelties of this paper include:

1. the notion of *learning context* and it use to compare learning algorithms or to tune their performance.
2. the definition of *performance maps* and how they can be used to compare learners
3. the description of how to create approximate (partial) performance maps with relatively low computational cost yet providing 'satisfactory' information
4. the suggestion that previous research in the literature, has been implicitly using a weak version of the performance maps method, here described, usually performed informally by the authors before selecting the configuration to use in the learners discussed in their papers
5. the suggestion that comparison tables among learners, presented in the literature, would benefit from being expanded and recalculated according to performance maps to provide more insights to the reader looking for the best learner/configuration when dealing with a specific data set.
6. the observation that performance maps fit nicely in the scope of the No Free Lunch Theorem (NFL) [36]. The NFL theorem states that no learning algorithms can outperform all the others over all data sets. Our proposal makes explicit that changing parameter settings of a learning algorithm produces a different learner which usually has different a performance.

The paper is organized as follows: in Sect. 2, we summarize the standard procedure to compare learning systems, in Sect. 3, we introduce the learning systems and the meta-optimizers used in the study, in Sect. 4 and 5, we discuss their parameter spaces, Sect. 6 describes the data sets used in the experiments, Sect. 7 reports the experimental study, and finally some conclusions close the paper.

2   State of the Art in Comparing Learning Algorithms

The standard procedure to compare learning algorithms consists in contrasting their performances on several data sets. It must be added that the comparison is done after an ad hoc selection of the better performing parameter settings for the learners. Usually manually discovered by running some trial tests.

Traditional performance measures include: accuracy, error rate, R, etc. Their values are generally determined by using a statistical methodology called n-fold cross validation (usually 5 or 10 folds are selected) on the whole available data in order to determine a performance interval (mean $\pm$ standard deviation) with known statistical confidence [32, 35].

Because performance measures reduce to a single value the whole learner's behavior, they may potentially miss important aspects of the underlying learning process like, for instance, the distribution of the performances over the parameter space of the learner.

In addition to traditional performance measures, other methodologies exist to evaluate a learner's performance. For instance: the Area Under the ROC Curve (AUC) [3] or the rolling cross validation [1,18,31]. AUC is applicable to any classifier producing a score for each case, but less appropriate for discrete classifiers like decision trees. Rolling cross validation is only applicable to specific data types like time series or data streams [1,18,31]. In fact, more recent performance measures are not generally applicable across learners or data types.

We then believe that, when learners need to be compared, the information provided by the above performance measures could be enhanced by including some insights about the distribution of performances on the learners' parameter spaces. The latter information would allow, for instance, to take into account the probability of achieving a high performance by randomly selecting a parameter from the learner's parameter space with uniform probability. Thus providing a measure of confidence or stability in the best performance achieved in the learning context under study.

2.1 Our Proposal: Comparing Learning Algorithms with Performances Maps and Their HP(k) Values

This study proposes to compare learning algorithms by confronting their *performances maps* and their HP(k) values. As said, given a learning context LC, its performance map $Pmap(LC)$ is the collection of pairs ¡ s, $L(s)$¿, with s in $MOPS$, and $L(s)$ as the performance of L run with settings s. From $Pmap(LC)$, it is very simple to determine its best performance $best(LC)$ (the map's maximum).

The High Performance function of a map $HP_{Pmap(LC)}(k)$ is defined as the ratio between the number of parameter settings in $MOPS$ producing a performance with distance k from $best(LC)$, and the cardinality of $MOPS$, as in Eq. (1).

$$HP_{Pmap(LC)}(k) = \frac{|\{p|p \in MOPS \text{ and } L(p) >= best(LC) * (1 - k)\}|}{|MOPS|} \quad (1)$$

where $p \in MOPS, 0 < k < 1$, and $L(p)$ is the performance observed by running L with parameter settings p on the data D. In the following, we will use $HP_{LC}(k)$, or simply $HP(k)$ when the learning context is clear, as shorthand for $HP_{Pmap(LC)}(k)$.

$HP_{LC}(k)$ also represents the fraction of the map area above a certain performance level $(best(LC) * (1 - k))$ over the whole map extension. And, from another point of view, $HP_{LC}(k)$ is an estimate of the cumulative distribution function $Prob_{LC}(X > best(LC) * (1 - k))$, where X is $L(s)$ and s is randomly taken from $MOPS$ with uniform distribution.

We will show, in the experimental session, the values of $HP_{LC}(k)$ for several learning contexts.

3   Learners and Meta Optimization Methods

As said, the aim of our work is to compare learners across learning contexts by using performance maps. In order to practically show how our proposal works, we selected two learners and two meta-optimization methods so that we were able to present full set of experiments.

Decision Trees (DT) [30] and Support Vector Machines (SVM) [6] are selected as learners because they internally represent knowledge in very different way, thus demonstrating the general applicability of our methodology. And as meta-optimization methods, we selected Grid Search, which consists in the exhaustive enumeration of a input parameter space, and Simple Genetic Algorithm (SGA) [10,14], in order to account for the case of partial search of the input parameter space, and the ensuing partial performance map. We note that one can choose to build a partial performance map as it has a lower computational cost than a complete one.

4   The Parameter Spaces for the Selected Learners

The chosen parameter spaces for DT and SVM are shown in Tables 1 and 2. These are the parameter spaces searched by the meta-optimizer.

In the case of DT, the parameters that mostly affects its results have been identified in: minimum impurity decrease (decrease of a node's impurity to allow for a node splitting), minimum samples (the minimum number of samples required to split an internal node), and max depth (the maximum allowed depth of the tree). The parameter space for DT contains combinations of values for the three selected parameters. Similarly, for SVM, the chosen parameters are gamma, kernel, and C value, which affect the types of hyperplanes to be used and their boundary positions (margin distance). Again the combination of values for these three parameters define the parameter space for SVM.

It is important to note that our methodology is not limited by the number of parameters used to define a parameter space. In this experimentation, we define the parameter spaces with only three parameters per learner simply because this choice will allow to draw 3-dimensional representation of the performance maps build in the experiments. Thus facilitating the understanding of our work. If we had used more parameters it would have been difficult to show the results in a graphical form.

Table 1. Value ranges for the selected parameters of DT.

Learner	Min Impurity	Min Samples	Max Depth	Timeout (secs)
DT	{i/10 for i = 0 to 6}	{i for i = 2 to 150 step 10}	{i for i = 1 to 160 step 10}	40

Table 2. Value ranges for the selected parameters of SVM.

Learner	Gamma	Kernel	C value	Timeout (secs)
SVM	scale auto	linear poly rbf sigmoid	{i/100 for i = 1 to 200 step 20} ∪ {i for i = 2 to 200 step 20}	40

The Timeout columns in the tables report the maximum number of seconds an experiment will run before timing out. As an anticipation, an experiment consists in performing several 10 fold cross validations of the selected learner on the available data in order to meta-optimize it.

5   Parameter Settings for the Meta-Optimization Methods

In the case of Grid Search, no parameters affects its behavior because all points in the given parameter space are evaluated.

In the case of SGA, instead, it is known that the population size and the maximum number of generations can deeply affect the result found by a genetic algorithm. Here is why, in order to find the best parameter settings for the SGA, we meta-optimized the SGA by using a Grid Search applied to the following parameter ranges: population size (30, 50, 80), max number of generations (30, 50, 80), crossover probability (0.5, 0.7, 0.9), and learner (DT or SVM).

As performance measure, we were interested in the genetic algorithm discovering a parameter settings performing as close as possible to the best performance discovered by Grid Search when used as a meta-optimizer in the learning contexts. Also by using the lowest possible population size and max generations.

The found parameter settings for SGA are: population size equal to 50, max number of generations equal to 50, and crossover probability equal to 0.9. The fact that genetic algorithms, in general, are robust learners makes it quite easy to find one of the many suitable parameter settings [14, 15].

We kept the remaining parameters of SGA to their default values as set in the python library GeneticAlgorithm (https://pypi.org/project/geneticalgorithm/) from which we built the SGA used in this study.

6   Data Set Descriptions

To perform the experiments in our study, we selected four data sets with varying characteristics from the UCI Machine Learning repository:

1. Mushrooms - 8124 instances, 22 attributes (categorical), classification task: to predict if a mushrooms is either edible or poisonous from some physical characteristics [34].
2. Congressional Voting Records - 435 instances, 16 attributes (categorical), classification task: predicting Republican or Democratic membership from vote record [34].

An open research question is if the proposed methodology needs to be extended when different data types like for instance financial time series [16,18,19,24] or unusual domains are considered [9].

7   Experimental Analysis

As experimental platform, we implemented the code in Python 3.8, making use of SciKit Learn [29], and used a Dell XPS 13, with Intel CPU I7, 7th gen, and 16 GB RAM, as hardware. We used the implementation of DT and SVM as provided in python's SciKit Learn library, and directly implemented the meta-optimization algorithms: Grid search and SGA.

Moreover, we remind that the following empirical results are to be considered the initial phase of a multi-year long research line where more learning contexts will be studied. Given a learning context LC, an experiment consists in using the meta-optimizer M to find the best performing parameter settings for the learner L. Each parameter settings evaluated by M requires performing a 10 fold cross validation in order to ensure the correct measurement of L's performance.

In Table 3, all the experiments performed are reported with their best performances across the 16 learning contexts considered. Performances are measured with the accuracy measure for classification tasks, and with the coefficient of determination R^2 for the regression task (Abalone data set). The time column shows the time to run a complete experiment. The following findings appear from Table 3:

1. some learning contexts do not admit for a perfect solution;
2. learning contexts with Grid Search, as a meta-optimizer, usually takes longer than SGA. This is reasonable because Grid Search has to evaluate all settings for L, whereas SGA will consider only some of them.
3. learning contexts with DT usually run in less time than those exploiting SVM.

The structure of these findings follows the standard used in the literature to assess learning systems.

In this novel research line, we aim to propose to the research community to augment the way machine learning systems are compared by including also information from *performance maps* and their *high performance values*.

7.1   Performance Maps

We recall that a performance map $Pmap(LC)$ for a learning context LC is the set of pairs $< s, L(s) >$, where s is a parameter settings in $MOPS$ and $L(s)$ is the performance obtained by running L with settings s.

Because a performance map, for a learning context, shows the distribution of performances for the associated learner over (part of) its parameter space, it then provides information about how frequent high performing parameter settings

Table 3. Meta-optimization of learners in several learning contexts.

Data set	Learner and Meta Optimization	Best Accuracy/R^2	Std	Evaluated points	Time
Mushrooms	DT - Grid	1.0	0.0	1440	197.45
	DT - SGA	1.0	0.0	49	6.70
	SVM - Grid	1.0	0.0	160	1000.25
	SVM - SGA	1.0	0.0	47	320.30
Congr. Votes	DT - Grid	0.96	0.03	1440	18.08
	DT - SGA	0.96	0.03	272	5.28
	SVM - Grid	0.97	0.02	160	5.50
	SVM - SGA	0.96	0.02	129	6.11

are. Also it shows the specific value ranges for those high performing settings. Thus, in addition, a performance map provides an insight about how robust the associated learner is to changes to its settings.

Building a performance map then could be particularly useful when selecting a learner for some novel data, because it provides information on the robustness of the learner when different configurations are used, a situation which is bound to happen in real world usage of a learning system.

Here is why we believe that comparing learner by using performance maps provides more insights than the use of a single valued performance measure as traditionally done in the literature.

Figures 1 and 2 show the *performance maps* for the learning contexts of Table 3[1]. They peruse makes explicit that:

1. if we consider all learning contexts, DT performs better in a region of the parameter space where 'min impurity' is close to 0, 'min sample' is below 50 and 'max depth' is above 20. When increasing the 'min impurity' value above 0.2, the performance decreases abruptly and significantly
2. if we consider all learning contexts, SVM performs better in a region or the parameter space where 'gamma' is equal to 'scale', 'C-value' is lower than 1.0, and 'kernel' is 'poly', 'rbf' or 'linear'
3. however, if we are interested in a specific learner and data, the performance map shows the locations of the highest performing parameter settings and it displays how these regions varies in location and extensions across the parameter space

---

[1] We projected two parameters on the X axis for creating the 3-d graphs. In particular, we projected 'min impurity' and 'min samples' on the X axis and 'max depth' on the Y axis for DT. Then the label '0.1 - 20' on the X axis has to be interpreted as 'min impurity' = 0.2 and 'min samples = 20'. Instead, for SVM, we projected 'gamma' and 'C value' on the X axis and 'kernel' on the Y axis.

4. performance maps do not need to be complete to be useful. Completeness may require a high computational cost to achieve. Indeed, even partial performance maps are very helpful in selecting high performing parameter settings over just a blind selection of the same done by manually undertaking trial runs. Comparing performance maps using Grid Search with those using SGA demonstrates the point.

Moreover, by perusing the results in Table 3 and the performance maps, one can observe that even with relatively low computational costs, it is already possible to find high performing parameter settings when an effective meta-optimizer, such as SGA, is applied to explore the learner's parameter space.

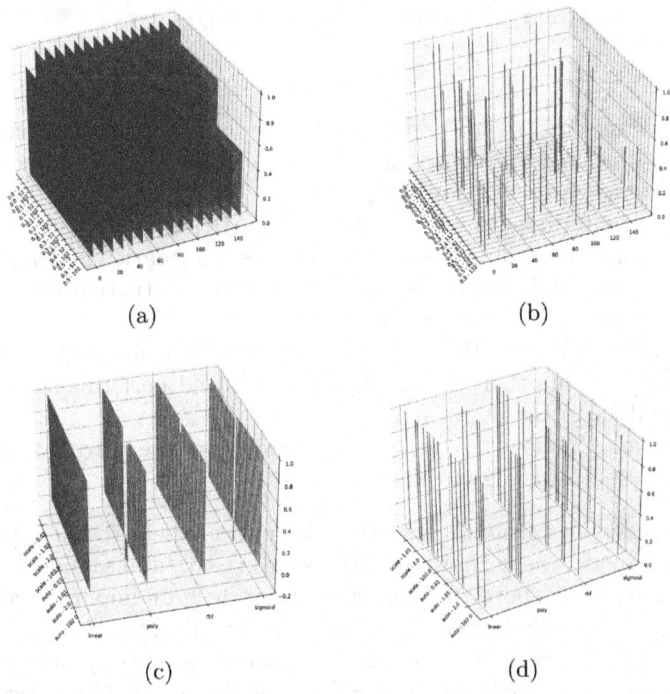

(a)   (b)

(c)   (d)

Fig. 1. Performance maps for the mushrooms data set. In the cases of (DT, Grid search) (a), (DT, SGA) (b), (SVM, Grid search) (c), and (SVM, SGA) (d).

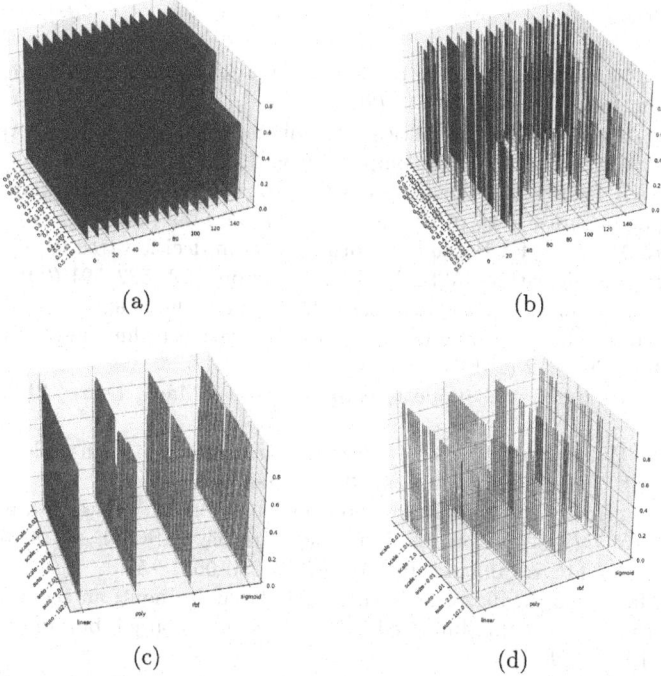

Fig. 2. Performance maps for the congressional voting records data set. In the cases of (DT, Grid search) (a), (DT, SGA) (b), (SVM, Grid search) (c), and (SVM, SGA) (d).

8   Conclusions

The study describes a novel methology to map learning algorithms on data (performance map) in order to gain more insights in the distribution of their performances across their parameter space. The initial research findings of a multi-year long research activity are discussed. It appears that this approach provides useful information when selecting the best configuration for a learning context and when comparing alternative learners. To formalize the above ideas, we introduced the notions of learning context, performance map, and high performance function. We then applied the concepts to a variety of learning contexts to show their capabilities.

The initial findings support the hypothesis that the proposed methodology can provide more information on the robustness of a learner in a given learning context thus enriching the traditional single-valued performance measures used in literature when comparing learners.

Future research directions are: to study the application of this methodology to more sophisticated learning systems such as agent based systems for modeling complex time series in financial applications [17,20,23,25–28] or to understand algorithms performance with respect to the explainability of their outcomes [21, 22] or to observe what will happen when neural networks are used as learners perhaps control applications [13].

References

1. Bergmeir, C., Benítez, J.M.: On the use of cross-validation for time series predictor evaluation. Inf. Sci. **191**, 192–213 (2012)
2. Blum, C., Roli, A.: Metaheuristics in combinatorial optimization: overview and conceptual comparison. ACM Comput. Surv. **35**(3), 268–308 (2003)
3. Bradley, A.P.: The use of the area under the roc curve in the evaluation of machine learning algorithms. Pattern Recogn. **30**(7), 1145–1159 (1997)
4. Camilleri, M., Neri, F.: Parameter optimization in decision tree learning by using simple genetic algorithms. WSEAS Trans. Comput. **13**, 582–591 (2014)
5. Camilleri, M., Neri, F., Papoutsidakis, M.: An algorithmic approach to parameter selection in machine learning using meta-optimization techniques. WSEAS Trans. Syst. **13**(1), 203–212 (2014)
6. Cortes, C., Vapnik, V.: Support-vector networks. Mach. Learn. **20**(3), 273–297 (1995)
7. Eiben, A., Hinterding, R., Michalewicz, Z.: Parameter control in evolutionary algorithms. IEEE Trans. Evol. Comput. **3**(2), 124–141 (1999)
8. Feurer, M., Hutter, F.: Hyperparameter optimization. In: Hutter, F., Kotthoff, L., Vanschoren, J. (eds.) Automated Machine Learning. TSSCML, pp. 3–33. Springer, Cham (2019). https://doi.org/10.1007/978-3-030-05318-5_1
9. García-Margariño, I., Plaza, I., Neri, F.: ABS-mindburnout: an agent-based simulator of the effects of mindfulness-based interventions on job burnout. J. Comput. Sci. **36**, 101012 (2019)
10. Goldberg, D.: Genetic Algorithms in Search, Optimization, and Machine Learning. Addison-Wesley, Reading, Ma (1989)
11. Grefenstette, J.J.: Optimization of control parameters for genetic algorithms. IEEE Trans. Syst. Man Cybern. **16**(1), 122–128 (1986)
12. Lorenzo, P.R., Nalepa, J., Kawulok, M., Ramos, L.S., Pastor, J.R.: Particle swarm optimization for hyper-parameter selection in deep neural networks. In: Proceedings of the Genetic and Evolutionary Computation Conference, pp. 481–488. GECCO 2017. ACM (2017)
13. Marino, A., Neri, F.: PID tuning with neural networks. In: Nguyen, N.T., Gaol, F.L., Hong, T.-P., Trawiński, B. (eds.) ACIIDS 2019. LNCS (LNAI), vol. 11431, pp. 476–487. Springer, Cham (2019). https://doi.org/10.1007/978-3-030-14799-0_41
14. Neri, F.: Traffic packet based intrusion detection: decision trees and genetic based learning evaluation. WSEAS Trans. Comput. **4**(9), 1017–1024 (2005)
15. Neri, F.: PIRR: a methodology for distributed network management in mobile networks. WSEAS Trans. Inf. Sci. Appl. **5**(3), 306–311 (2008)
16. Neri, F.: Learning and predicting financial time series by combining natural computation and agent simulation. In: Di Chio, C., et al. (eds.) EvoApplications 2011. LNCS, vol. 6625, pp. 111–119. Springer, Heidelberg (2011). https://doi.org/10.1007/978-3-642-20520-0_12
17. Neri, F.: Agent-based modeling under partial and full knowledge learning settings to simulate financial markets. AI Commun. **25**(4), 295–304 (2012)
18. Neri, F.: A comparative study of a financial agent based simulator across learning scenarios. In: Cao, L., Bazzan, A.L.C., Symeonidis, A.L., Gorodetsky, V.I., Weiss, G., Yu, P.S. (eds.) ADMI 2011. LNCS (LNAI), vol. 7103, pp. 86–97. Springer, Heidelberg (2012). https://doi.org/10.1007/978-3-642-27609-5_7
19. Neri, F.: Learning predictive models for financial time series by using agent based simulations. In: Nguyen, N.T. (ed.) Transactions on Computational Collective

Intelligence VI. LNCS, vol. 7190, pp. 202–221. Springer, Heidelberg (2012). https://doi.org/10.1007/978-3-642-29356-6_10

20. Neri, F.: Coevolution and learning symbolic concepts: statistical validation: empirical statistical validation of co-evolutive machine learning systems. In: Neri, F. (ed.) 7th International Conference on Machine Learning Technologies ICMLT 2022, pp. 244–248. ACM (2022)

21. Neri, F.: Explainability and interpretability in agent based modelling to approximate market indexes. In: Neri, F. (ed.) 8th International Conference on Machine Learning Technologies ICMLT 2023, in press. ACM (2023)

22. Neri, F.: Explainability and interpretability in decision trees and agent based modelling when approximating financial time series, a matter of balance with performance. In: Neri, F. (ed.) 2023 8th International Conference on Computational Intelligence and Applications ICCIA 2023, in press. Springer press (2023)

23. Neri, F., Margariño, I.: Simulating and modeling the DAX index and the USO Etf financial time series by using a simple agent-based learning architecture. Expert Syst. **37**(4), 12516 (2020)

24. Neri, F.: Software agents as a versatile simulation tool to model complex systems. WSEAS Trans. Info. Sci. and App. **7**(5), 609–618 (2010)

25. Neri, F.: Case study on modeling the silver and nasdaq financial time series with simulated annealing. In: Rocha, Á., Adeli, H., Reis, L.P., Costanzo, S. (eds.) WorldCIST'18 2018. AISC, vol. 746, pp. 755–763. Springer, Cham (2018). https://doi.org/10.1007/978-3-319-77712-2_71

26. Neri, F.: Combining machine learning and agent based modeling for gold price prediction. In: Cagnoni, S., Mordonini, M., Pecori, R., Roli, A., Villani, M. (eds.) WIVACE 2018. CCIS, vol. 900, pp. 91–100. Springer, Cham (2019). https://doi.org/10.1007/978-3-030-21733-4_7

27. Neri, F.: How to identify investor's types in real financial markets by means of agent based simulation. In: 6th International Conference on Machine Learning Technologies ICMLT 2021, pp. 144–149. ACM (2021)

28. Neri, F.: Unpublished result: Domain specific concept drift detectors for predicting financial time series. https://arxiv.org/abs/2103.14079. Arxiv.org (UP)

29. Pedregosa, F., et al.: Scikit-learn: machine learning in Python. J. Mach. Learn. Res. **12**, 2825–2830 (2011)

30. Quinlan, J.R.: C4.5: programs for machine learning. Morgan Kaufmann, California (1993)

31. Racine, J.: Consistent cross-validatory model-selection for dependent data: hv-block cross-validation. J. Econometr. **99**(1), 39–61 (2000)

32. Refaeilzadeh, P., Tang, L., Liu, H.: Cross-Validation, pp. 532–538. Springer, US, Boston, MA (2009). https://doi.org/10.1007/978-0-387-39940-9_565

33. Reif, M., Shafait, F., Dengel, A.: Meta-learning for evolutionary parameter optimization of classifiers. Mach. Learn. **87**(3), 357–380 (2012)

34. Schlimmer, J.C.: Concept acquisition through representational adjustment. Doctoral dissertation, Department of Information and Computer Science, University of California, Irvine, CA (1987)

35. Stone, M.: Cross-validatory choice and assessment of statistical predictions. discussion. J. Royal Statist. Soc. Ser. **B 36**, 111–147 (1974)

36. Wolpert, D., Macready, W.: No free lunch theorems for optimization. IEEE Trans. Evol. Comput. **1**(1), 67–82 (1997)

Author Index

© The Editor(s) (if applicable) and The Author(s), under exclusive license
to Springer Nature Switzerland AG 2023
F. Neri et al. (Eds.): CCCE 2023, CCIS 1823, p. 233, 2023.
https://doi.org/10.1007/978-3-031-35299-7

Printed in the United States
by Baker & Taylor Publisher Services